MECHANISMS IN PLANT DEVELOPMENT

Mechanisms in Plant Development

Ottoline Leyser
Department of Biology
Box 373
University of York
YO10 5YW

and

Stephen Day
28 St Oswalds Road
York
YO10 4PF

Blackwell
Publishing

BLACKWELL PUBLISHING
350 Main Street, Malden, MA 02148-5020, USA
9600 Garsington Road, Oxford OX4 2DQ, UK
550 Swanston Street, Carlton, Victoria 3053, Australia

First published 2003 by Blackwell Publishing Ltd

4   2007

Library of Congress Cataloging-in-Publication Data has been applied for.

ISBN-13: 978-0-86542-742-6 (paperback)

A catalogue record for this title is available from the British Library.

Set in 10/12 pt Galliard
by Kolam Information Services Pvt. Ltd, Pondicherry, India

The publisher's policy is to use permanent paper from mills that operate a sustainable
forestry policy, and which has been manufactured from pulp processed using
acid-free and elementary chlorine-free practices. Furthermore, the publisher ensures
that the text paper and cover board used have met acceptable environmental
accreditation standards.

For further information on
Blackwell Publishing, visit our website:
www.blackwellpublishing.com

Contents

Chapter 5: Axis development in the leaf and flower, 74

Chapter 6: Position relative to a particular cell, tissue or organ, 110

Chapter 7: Light, 138

Chapter 8: Environmental information other than light, 165

Chapter 9: The coordination of development, 190

Preface

Developmental biology is the study of how each cell in a multicellular organism acquires and maintains its specialized function. In plants, development is continuous, occurring throughout the life cycle. Multiple factors, both environmental and endogenous, combine to regulate cell specification, generating the enormous diversity of plant forms.

This book is about the mechanisms that regulate plant development. It is structured around these mechanisms and not around the stages of the life cycle, because similar regulatory mechanisms act at different stages of the life cycle and in different parts of the plant. The book is intended for final-year undergraduate courses in plant development and for graduate readers. It is obviously not a comprehensive treatise of all that is known about plant development, but we hope it will provide a conceptual framework from which to build an understanding of the subject.

We would like to thank Francesca and Joshua for going to bed on time every night so that we could write it.

Introduction

The central question of developmental biology is how does a single cell become a complex organism. The initial answer to this question must be descriptive. Reasonably complete cellular-level accounts of development now exist for a number of species, cataloguing the combinations of cell division, cell growth, cell differentiation, cell death and—in animals—cell migration that generate the adult organism. However, whilst a good description is essential, in order to understand development we must understand the mechanisms that control it. What factors control the behaviour of cells, directing them toward division, growth, migration, differentiation or death?

This book discusses the mechanisms underlying the development of flowering plants (the angiosperms). Plant development is not restricted to any one phase of the life cycle, but rather new structures such as leaves, roots and flowers are produced continually. Research aimed at understanding the control of plant development must therefore cover all stages of the plant's life cycle. None the less, similar developmental mechanisms may operate at different stages and consequently this book is not organized around the plant's life cycle, but instead it focuses on the developmental mechanisms themselves.

To put the discussion of mechanism into context, Chapter 1 gives a brief description of angiosperm development. This is intended both as an introduction for those who are unfamiliar with plant development, and as a reference to accompany later chapters. Chapter 2 then considers the implications of key cellular and larger scale characteristics of plant development for the study of development mechanisms.

To produce a functional plant, cells must adopt fates appropriate to their position. Leaf cells must adopt leaf fates; root cells must adopt root fates; cells on the surface of the leaf must behave differently to cells in internal layers; and so on. Chapters 3, 4, 5 and 6 discuss how cell fate is related to the position of the cell in the plant. Chapter 3 considers cell-intrinsic information, such as lineage. Chapters 4, 5 and 6 describe research into the mechanisms that generate cell-extrinsic positional information.

One of the most striking characteristics of plant development is its plasticity in response to environmental cues. This encompasses both spatial aspects of development, such as the direction of growth; and temporal aspects such as the timing of bud growth and the production of flowers. Chapters 7 and 8 discuss developmental responses to environmental information. Chapter 7 considers the many developmental effects of light. Chapter 8 outlines research into responses to other environmental cues. The effects of internal and environmental information on development are closely coordinated. Chapter 9 discusses this topic in relation to the development of the shoot.

Last of all, Chapter 10 compares the mechanisms that control plant development with those that operate in animals.

Sources for figures

Aida, M., Ishida, T. & Tasaka, M. (1999) Shoot apical meristem and cotyledon formation during *Arabidopsis* embryogenesis: interaction among the *CUP-SHAPED COTYLEDON* and *SHOOT MERISTEMLESS* genes. *Development* **126**, 1563–1570. [Fig. 4.10]

Barton, M.K. & Poethig, R.S. (1993) Formation of the shoot apical meristem in *Arabidopsis thaliana*: an analysis of development in the wild type and in the *shoot meristemless* mutant. *Development* **119**, 823–831. [Fig. 9.3]

Berger, F., Taylor, A. & Brownlee, C. (1994) Cell fate determination by the cell wall in early *Fucus* development. *Science* **263**, 1421–1423. [Fig. 4.3]

Bowman, J.L. & Eshed, Y. (2000) Formation and maintenance of the shoot apical meristem. *Trends in Plant Science* **5**, 110–115. [Fig. 9.5a]

Bradley, D., Carpenter, R., Copsey, L. *et al.* (1996) Control of inflorescence architecture in *Antirrhinum*. *Nature* **379**, 791–797. [Fig. 5.17a]

Chory, J. (1997) Light modulation of vegetative development. *Plant Cell* **9**, 1225–1234. [Fig. 7.1a]

Clark, S.E., Running, M.P. & Meyerowitz, E.M. (1995) *CLAVATA3* is a specific regulator of shoot and floral meristem development affecting the same processes as *CLAVATA1*. *Development* **121**, 2057–2067. [Fig. 9.3]

Cleary, A.L. & Smith, L.G. (1998) The *Tangled1* gene is required for spatial control of cytoskeletal arrays associated with cell division during maize leaf development. *The Plant Cell* **10**, 1875–1888. [Fig. 3.7]

Coen, E.S. (1996) Floral symmetry. *EMBO Journal* **15**, 6777–6788. [Fig. 5.16]

Di Laurenzio, L., Wysocka-Diller, J., Malamy, J.E. *et al.* (1996) The *SCARECROW* gene regulates an asymmetric cell division that is essential for generating the radial organization of the *Arabidopsis* root. *Cell* **86**, 423–433. [Fig. 4.11]

Dolan, L., Janmaat, K., Willemsen, V. *et al.* (1993) Cellular organisation of the *Arabidopsis thaliana* root. *Development* **119**, 71–84. [Fig. 6.4b]

Fleming, A.J., McQueen-Mason, S., Mandel, T. & Kuhlemeier, C. (1997) Induction of leaf primordia by the cell wall protein expansin. *Science* **276**, 1415–1418. [Fig. 6.12]

Foard, D.E. (1971) The initial protrusion of a leaf primordium can form without concurrent periclinal cell divisions. *Canadian Journal of Botany* **49**, 1601–1603. [Fig. 3.8]

Freeling, M. (1992) A conceptual framework for maize leaf development. *Developmental Biology* **153**, 44–58. [Fig. 5.8]

Fukaki, H., Wysocka-Diller, J., Kato, T. *et al.* (1998) Genetic evidence that the endodermis is essential for shoot gravitropism in *Arabidopsis thaliana*. *Plant Journal* **14**, 425–430. [Fig. 8.3]

Hadfi, K., Speth, V. & Neuhaus, G. (1998) Auxin-induced developmental patterns in *Brassica juncea* embryos. *Development* **125**, 879–887. [Fig. 4.8]

Hardtke, C.S. & Berleth, T. (1998) The *Arabidopsis* gene *MONOPTEROS* encodes a transcription factor mediating embryo axis formation and vascular development. *EMBO Journal* **17**, 1405–1411. [Fig. 4.7]

Hareven, D., Gutfinger, T., Parnis, A., Eshed, Y. & Lifschitz, E. (1996) The making of a compound leaf: genetic manipulation of leaf architecture in tomato. *Cell* **84**, 735–744. [Fig. 5.10b]

Haughn, G.W., Schultz, E.A. & Martinez-Zapater, J.M. (1995) The regulation of flowering in *Arabidopsis thaliana*: meristems, morphogenesis, and mutants. *Canadian Journal of Botany* **73**, 959–981. [Fig. 9.9]

Helariutta, Y., Fukaki, H., Wysocka-Diller, J. *et al.* (2000) The *SHORT-ROOT* gene controls radial patterning of the *Arabidopsis* root through radial signaling. *Cell* **101**, 555–567. [Fig. 4.11]

Hempel, F.D., Welch, D.R. & Feldman, L.J. (2000) Floral induction and determination: where is flowering controlled? *Trends in Plant Science* **5**, 17–21. [Fig. 9.10]

Kropf, D.L., Bisgrove, S.R. & Hable, W.E. (1999) Establishing a growth axis in fucoid algae. *Trends in Plant Science* **4**, 490–494. [Figs 4.1, 4.2]

Laux, T., Mayer, K.F.X., Berger, J. & Jürgens, G. (1996) The *WUSCHEL* gene is required for shoot and floral meristem integrity in *Arabidopsis. Development* **122**, 87–96. [Fig. 9.3]

Lee, M.M. & Schiefelbein, J. (1999) WEREWOLF, a MYB-related protein in *Arabidopsis*, is a position-dependent regulator of epidermal cell patterning. *Cell* **99**, 473–483. [Figs 6.4b, 6.6]

Liljegren, S.J., Gustafson-Brown, C., Pinyopich, A., Ditta, G.S. & Yanofsky, M.F. (1999) Interactions among *APETALA1, LEAFY,* and *TERMINAL FLOWER1* specify meristem fate. *Plant Cell* **11**, 1007–1018. [Fig. 9.9]

Lotan, T., Ohto, M., Yee, K.M. *et al.* (1998) *Arabidopsis* LEAFY COTYLEDON1 is sufficient to induce embryo development in vegetative cells. *Cell* **93**, 1195–1205. [Fig. 9.6d]

Lucas, W., Ding, B. & Van Der Schoot, C. (1993) Plasmodesmata and the supracellular nature of plants. *New Phytologist* **125**, 435–476. [Fig. 2.2]

Luo, D., Carpenter, R., Vincent, C., Copsey, L. & Coen, E. (1996) Origin of floral asymmetry in *Antirrhinum. Nature* **383**, 794–799. [Fig. 5.17b]

McHale, N.A. (1992) A nuclear mutation blocking initiation of the lamina in leaves of *Nicotiana sylvestris. Planta* **186**, 355–360. [Fig. 5.5]

McHale, N.A. & Marcotrigiano, M. (1998) *LAM1* is required for dosoventrality and lateral growth of the leaf blade in *Nicotiana. Development* **125**, 4235–4243. [Fig. 5.5]

Mattheck, C. (1990) Why they grow, how they grow: the mechanics of trees. *Arboricultural Journal* **14**, 1–17 [Fig. 8.1]

Mayer, U., Büttner, G. & Jürgens, G. (1993) Apical–basal pattern formation in the *Arabidopsis* embryo: studies on the role of the *gnom* gene. *Development* **117**, 149–162. [Fig. 4.4]

Meinke, D.W. (1992) A homoeotic mutant of *Arabidopsis thaliana* with leafy cotyledons. *Science* **258**, 1647–1650. [Fig. 9.6b]

Ogas, J., Cheng, J-C., Sung, Z.R. & Somerville, C. (1997) Cellular differentiation regulated by gibberellin in the *Arabidopsis thaliana pickle* mutant. *Science* **277**, 91–94. [Fig. 9.6c]

Parcy, F., Nilsson, O., Busch, M.A., Lee, I. & Weigel, D. (1998) A genetic framework for floral patterning. *Nature* **395**, 561–566. [Fig. 5.12]

Poethig, R.S. & Sussex, I.M. (1985) The cellular parameters of leaf development in tobacco: a clonal analysis. *Planta* **165**, 170–184. [Fig. 3.1]

Przemeck, G.K.H., Mattsson, J., Hardtke, C.S., Sung, Z.R. & Berleth, T. (1996) Studies on the role of the *Arabidopsis* gene *MONOPTEROS* in vascular development and plant axialization. *Planta* **200**, 229–237. [Fig. 4.7]

Rijven, A.H.G.C. (1968) Randomness in the genesis of phyllotaxis. I. The initiation of the first leaf in some Trifolieae. *New Phytologist* **67**, 247–256. [Fig. 6.13]

Sabatini, S., Beis, D., Wolkenfelt, H. *et al.* (1999) An auxin-dependent distal organizer of pattern and polarity in the *Arabidopsis* root. *Cell* **99**, 463–472. [Fig. 4.9]

Sack, F.D. (1997) Plastids and gravitropic sensing. *Planta* **203**, S63–S68. [Fig. 8.2]

Salisbury, F.B. & Ross, C.W. (1992) *Plant Physiology,* 4th edn. Wadsworth Publishing Co., Belmont, CA. [Fig. 7.11]

Satina, S., Blakeslee, A.F. & Avery, A.G. (1940) Demonstration of the three germ layers in the shoot apex of *Datura* by means of induced polyploidy in periclinal chimeras. *American Journal of Botany* **27**, 895–905. [Fig. 3.6]

Scheres, B., Di Laurenzio, L., Willemsen, V. *et al.* (1995) Mutations affecting the radial organisation of the *Arabidopsis* root display specific defects thoughout the embryonic axis. *Development* **121**, 53–62. [Figs 3.9, 4.11]

Schnittger, A., Folkers, U., Schwab, B., Jürgens, G. & Hülskamp, M. (1999) Generation of a spacing pattern: the role of *TRIPTYCHON* in trichome patterning in *Arabidopsis. Plant Cell* **11**, 1105–1116. [Figs 6.1, 6.3]

Schoof, H., Lenhard, M., Haecker, A. *et al.* (2000) The stem cell population of *Arabidopsis* shoot meristems is maintained by a regulatory loop between the *CLAVATA* and *WUSCHEL* genes. *Cell* **100**, 635–644. [Fig. 9.5b]

Serrano, N. & O'Farrell, P. (1997) Limb morphogenesis: connections between patterning and growth. *Current Biology* 7, R186–R195. [Fig. 10.1]

Siegfried, K.R., Eshed, Y., Baum, S.F. *et al.* (1999) Members of the *YABBY* gene family specify abaxial cell fate in *Arabidopsis*. *Development* 126, 4117–4128. [Fig. 5.4b]

Sinha, N.R., Williams, R.E. & Hake, S. (1993) Overexpression of the maize homeobox gene, *KNOTTED-1*, causes a switch from determinate to indeterminate cell fates. *Genes and Development* 7, 787–795. [Fig. 5.10a]

Smith, H. (1994) Sensing the light environment: the functions of the phytochrome family. In: *Photomorphogenesis in Plants*, 2nd edn (ed. R.E. Kendrick & G.H.M. Kronenberg), pp. 377–416. Kluwer Academic Publishers, Dordrecht. [Fig. 7.8a]

Steeves, T.A. & Sussex, I.M. (1989) *Patterns in Plant Development*, 2nd edn. Cambridge University Press, Cambridge. [Figs 5.2, 6.9]

Steinmann, T., Geldner, N., Grebe, M. *et al.* (1999) Coordinated polar localization of auxin efflux carrier PIN1 by GNOM ARF GEF. *Science* 286, 316–318. [Fig. 4.6]

Stewart, R.N. (1978) Ontogeny of the primary body in chimeral forms of higher plants. In: *The Clonal Basis of Development* (ed. S. Subtelny & I.M. Sussex), pp. 131–160. Academic Press, New York. [Fig. 3.5]

Stewart, R.N., Semeniuk, P. & Dermen, H. (1974) Competition and accommodation between apical layers and their derivatives in the ontogeny of chimeral shoots of *Pelargonium X hortorum*. *American Journal of Botany* 61, 54–67. [Fig. 5.11]

Tabata, T., Schwartz, C., Gustavson, E., Ali, Z. & Kornberg, T.B. (1995) Creating a *Drosophila* wing *de novo*, the role of engrailed, and the compartment border hypothesis. *Development* 121, 3359–3369. [Fig. 10.1]

Telfer, A., Bollman, K.M. & Poethig, R.S. (1997) Phase change and the regulation of trichome distribution in *Arabidopsis thaliana*. *Development* 124, 645–654. [Fig. 3.10]

Torii, K.U. & Deng, X-W. (1997) The role of COP1 in light control of *Arabidopsis* seedling development. *Plant, Cell and Environment* 20, 728–733. [Fig. 7.7]

Torres-Ruiz, R.A. & Jürgens, G. (1994) Mutations in the *FASS* gene uncouple pattern formation and morphogenesis in *Arabidopsis* development. *Development* 120, 2967–2978. [Fig. 3.9]

Van den Berg, C., Willemsen, V., Hage, W., Weisbeek, P. & Scheres, B. (1995) Cell fate in the *Arabidopsis* root meristem determined by directional signalling. *Nature* 378, 62–65. [Figs 3.4, 4.12]

Veit, B., Briggs, S.P., Schmidt, R.J., Yanofsky, M.F. & Hake, S. (1998) Regulation of leaf initiation by the *terminal ear 1* gene of maize. *Nature* 393, 166–168. [Fig. 6.11]

Vroemen, C.W., Langeveld, S., Mayer, U. *et al.* (1996) Pattern formation in the *Arabidopsis* embryo revealed by position-specific lipid transfer gene expression. *Plant Cell* 8, 783–791. [Fig. 4.4]

Waites, R. & Hudson, A. (1995) *phantastica*: a gene required for dorsoventrality of leaves in *Antirrhinum majus*. *Development* 121, 2143–2154. [Figs 5.3, 5.6]

Wardlaw, C.W. (1949) Further experimental observations on the shoot apex of *Dryopteris aristata* Druce. *Philosophical Transactions of the Royal Society, Series B* 233, 415–452. [Fig. 6.9]

Waring, P.F. & Phillips, I.D.J. (1970) *The Control of Growth and Differentiation in Plants*. Pergamon Press, Oxford. [Fig. 6.14]

Weigel, D. & Meyerowitz, E.M. (1994) The ABCs of floral homeotic genes. *Cell* 78, 203–209. [Fig. 5.12]

Yadegari, R., de Paiva, G.R., Laux, T. *et al.* (1994) Cell differentiation and morphogenesis are uncoupled in *Arabidopsis raspberry* embryos. *Plant Cell* 6, 1713–1729. [Fig. 4.13]

Zhang, H. & Forde, B.G. (2000) Regulation of *Arabidopsis* root development by nitrate availability. *Journal of Experimental Botany* 51, 51–59. [Fig. 8.8]

Zhang, H., Jennings, A., Barlow, P.W. & Forde, B.G. (1999) Dual pathways for regulation of root branching by nitrate. *Proceedings of the National Academy of Science, USA* 96, 6529–6534. [Fig. 8.9]

CHAPTER 1

An introduction to flowering plants

The angiosperm life cycle alternates between an extensive **diploid phase** and a more restricted **haploid phase**. The diploid phase is what we recognize as a plant. Its development consists of: (i) **embryogenesis**; (ii) **germination**; (iii) **primary development**, in which shoots and roots elongate and branch; and (iv) **secondary development**, in which shoots and roots thicken. Overlapping primary and secondary development, plant development may be **vegetative** or **reproductive**. Vegetative development is divided further into a **juvenile phase** and an **adult phase**.

Alternation of generations

The names of the haploid and diploid phases of the plant's life cycle reflect the type of reproductive cells that each produces. The haploid phase of the plant is called the **gametophyte** because it produces male and female **gametes** by mitosis. These fuse to form the **zygote** from which the diploid phase develops. Likewise, the diploid phase of the plant is called the **sporophyte** because it produces haploid **spores** by meiosis, from which the haploid phase of the plant develops. In flowering plants there are separate male and female gametophytes, producing sperm and egg cells, respectively. The female gametophyte develops from a **megaspore** and the male gametophyte forms from a **microspore**. Both gametophytes consist of only a few cells that are entirely dependent on the sporophyte for their nutrition.

Gametophyte development

The female gametophyte, which is also called the **embryo sac**, develops within the **carpel** (Fig. 1.1). Carpels typically consist of an **ovary**, a filament called the **style** and a sticky receptacle called the **stigma**. The ovary

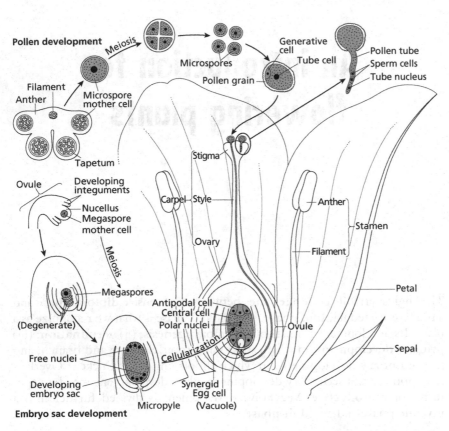

Pollen development

Meiosis

Microspores

Generative cell

Pollen grain

Tube cell

Pollen tube
Sperm cells
Tube nucleus

Filament
Anther

Microspore mother cell

Tapetum

Ovule

Developing integuments

Nucellus
Megaspore mother cell

Meiosis

Megaspores

(Degenerate)

Free nuclei

Developing embryo sac

Stigma

Carpel Style

Ovary

Antipodal cell
Central cell
Polar nuclei

Cellularization

Synergid
Egg cell
(Vacuole)

Micropyle

Anther

Stamen

Filament

Petal

Ovule

Sepal

Embryo sac development

Fig. 1.1 Gametophyte development in angiosperms. The female gametophyte develops in the ovule and forms the embryo sac. The male gametophyte develops in the anther and forms the pollen grain.

holds one or more ovules and these are the sites of embryo sac development. Each ovule contains a roughly egg-shaped mass of cells called the **nucellus** and surrounding this there are two outer layers of cells called the **integuments**. The integuments do not quite join at the tip of the ovule but leave a small gap called the **micropyle**. The development of the female gametophyte begins with the meiosis of a single cell in the nucellus just below the micropyle to create a strand of four haploid megaspores.

In most plants, three of the megaspores degenerate but the fourth, the one farthest from the micropyle, enlarges and undergoes three rounds of mitosis to create an embryo sac containing eight haploid nuclei distributed among seven cells. Three of the seven cells cluster around the end of the embryo sac closest to the micropyle. The cell in the middle of these three is the **egg cell** and the two cells flanking it are called **synergids**. At the other end of the embryo sac there is a group of three cells called the **antipodal cells**, and in the middle of the sac there is a single, binucleate **central cell**. The central cell's two nuclei are called the **polar nuclei**.

Male gametophytes develop in the **anther** at the top of the stamen. Here there are typically four pollen sacs, each of which contains a column of diploid **microspore mother cells** surrounded by a nutritive tissue called the **tapetum**. Each mother cell undergoes meiosis to form four haploid microspores and each of these develops into a pollen grain. Within the pollen grain, the microspore divides mitotically to produce a **tube cell** and a **generative cell**. In some species, the generative cell immediately divides again to give a pair of **sperm cells**. In most flowering plants, however, this division takes place later, in the tube that develops when a pollen grain germinates.

For fertilization to occur, the pollen grain must germinate on a compatible stigma. The pollen tube grows into the stigma, through the style to the ovary, and enters the ovule, normally through the micropyle. As the tube grows, the tube cell nucleus stays near to the tip and the two sperm cells follow behind. When the pollen tube reaches the embryo sac, it penetrates one of the synergids and then releases both sperm cells. When, or even before, this happens the synergids degenerate allowing the sperm cells access to both the egg cell and the central cell. One of the distinguishing features of angiosperms is that they have double fertilization. One sperm cell fuses with the egg cell to produce the zygote. The other sperm cell fuses with the two polar nuclei in the central cell to create a triploid nucleus from which the **endosperm** (a nutritive tissue) develops.

After fertilization has occurred, the remaining haploid cells of the embryo sac degenerate. Most parts of the parent flower also wither, with the exception of the ovaries which normally develop into the fruit. As the embryo grows, it destroys most of the nucellus. The two integuments, however, normally remain to form the coat of the new seed.

Development of the sporophyte

Embryogenesis

In most species, the zygote divides in a plane perpendicular to the long axis of the embryo sac to produce a large **basal cell** near to the micropyle, and a small **terminal cell** close to what was the central cell and is now the developing triploid endosperm (Fig. 1.2). Subsequent patterns of cell division are more variable, sometimes even within a species, but the appearance of the main tissues and organs of the embryo follows a predictable sequence. The first distinction to arise is between cells that will form the main body of the embryo and those that will produce the **suspensor**, a filament that connects the embryo to maternal tissue near the micropyle. It is always cells at the micropyle end of the embryo that form the suspensor, and cells nearer the centre of the embryo sac that develop into the embryo proper.

Embryos undergo a regular series of changes in shape. During early development, cell divisions occur with little or no increase in the embryo's total size, resulting in a ball of cells called the **globular embryo**. At this stage, different tissues appear. The outermost layer of cells forms an **epidermis** in which the cells are typically smaller than the cells in the underlying tissue. The epidermis of the embryo and the epidermis of immature regions

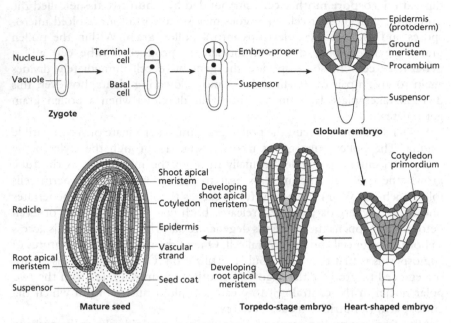

Fig. 1.2 Embryogenesis in *Arabidopsis thaliana*. The *Arabidopsis* embryo develops very rapidly. In many angiosperms each stage of embryo development is longer, resulting in larger embryos consisting of more cells.

of growing shoots and roots are sometimes called the **protoderm**. Deeper in the globular embryo, a distinction appears between relatively large cells with abundant vacuoles and a strand of smaller, less vacuolated cells. This strand will develop into **vascular tissue** and, in this immature state, it is called the **procambium**. The larger cells will form the **ground tissue** of the embryo and, in their immature state, they are sometimes called the **ground meristem**. The precursors of the root and shoot **apical meristems** appear at this stage, or later, as clusters of densely cytoplasmic cells at either end of the procambial strand. All post-embryonic cell lineages originate in the apical meristems. The root apical meristem forms at the end of the procambial strand nearest to the suspensor (and often incorporates the suspensor cell closest to the embryo proper). The shoot apical meristem develops away from the suspensor at the other end of the procambium.

The globular stage of embryo development ends when either one or two **cotyledons** begin to form near the site of the shoot apical meristem. In dicots, two cotyledons develop and the embryo changes from a globular embryo to a **heart-shaped embryo**, with the cotyledons as the two bulges at the top of the heart. The heart-shaped embryo elongates to form the **torpedo-stage embryo** while retaining the same pattern of tissues and organs. Monocots, of course, do not have a heart-shaped stage because they only form a single cotyledon. Depending on the relative sizes of the embryo and the seed, the embryo may fold over as it elongates.

The growth of most angiosperms pauses between the end of embryogenesis and the beginning of germination, but the extent of embryo development prior to this pause varies between species. In embryos with the most limited development, a shoot apical meristem and a root apical meristem form but immediately become dormant. In other species, some growth occurs at both apical meristems before dormancy: the shoot apical meristem initiates leaves to form the **plumule** and the root apical meristem initiates a short length of root called the **radicle**. In grasses, protective sheaths called the **coleoptile** and the **coleorhiza** enclose the plumule and the radicle, respectively.

Germination

Germination requires the correct combination of external cues such as moisture, temperature and light. During germination, extensive cell elongation forces the root out of the seed case and carries the shoot upwards. The exact anatomy of a seedling depends on the location of cell elongation in the shoot. There are two main possibilities and they are simplest to visualize for a dicot seedling (Fig. 1.3). If elongation takes place between the cotyledons and the radicle (in the **hypocotyl**), the cotyledons are lifted above ground

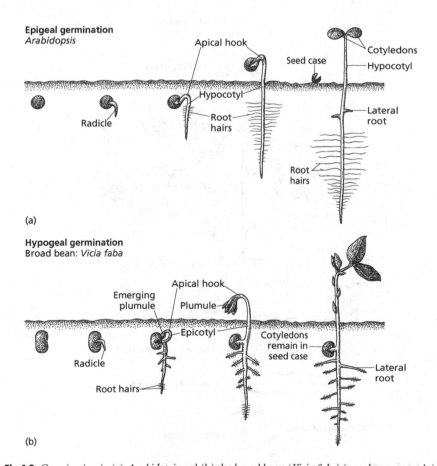

Fig.1.3 Germination in (a) *Arabidopsis* and (b) the broad bean (*Vicia faba*) (not drawn to scale).

and the seedling is called **epigeal**, meaning 'above the ground'. If elongation takes place between the cotyledons and the shoot apical meristem (in the **epicotyl**), the cotyledons remain underground and the seedling is called **hypogeal**, meaning 'below the ground'. In either case, the germinating shoot grows upwards in a hook shape so that it pulls rather than pushes the shoot tip through the soil. Once above the soil and in the light the hook straightens out, and the cotyledons or leaves of the plumule green up and expand.

Primary vegetative development

The primary development of the plant is that in which shoots and roots lengthen and branch. Shoots and roots elongate due to cell division and cell elongation in and immediately behind their apical meristems. Branching occurs by the development of additional, laterally placed meristems that become the apical meristems of the lateral shoots and roots. Figure 1.4 shows the structures produced during primary development.

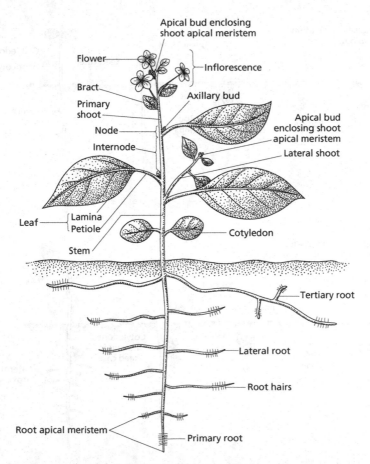

Fig. 1.4 Structures produced during primary development.

Primary development of roots

Root apical meristem

The root apical meristem (RAM) is not at the very tip of the root, but lies behind a protective shield called the **root cap** (Fig. 1.5). Cell divisions in the most apical layers of the meristem add cells to the root cap, replacing those worn away by friction between the growing root and the soil. Cell divisions deeper in the meristem contribute cells to the main tissues of the root. From the circumference inwards, these are the **epidermis**, the **cortex**, the **endodermis**, the **pericycle** and a central core of **vascular tissue**.

When the root is initiated, all cells in the apical meristem normally divide at about the same rate. In many species, however, cells in the centre of the meristem become inactive over time, forming a so-called **quiescent centre**. Cells in the quiescent centre do not lose the ability to divide rapidly but can be activated after damage to the apical meristem. In some plants, the quiescent centre also appears and disappears on a seasonal basis.

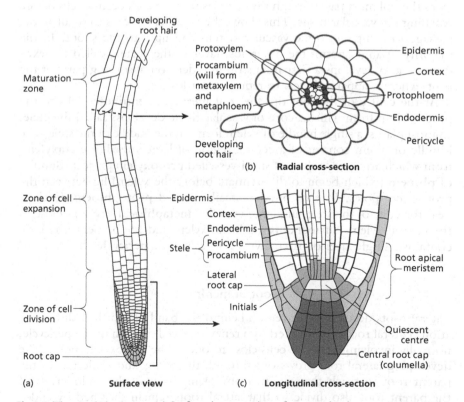

Fig. 1.5 The *Arabidopsis* root. (a) Surface view: a section of root cap has been omitted to show the epidermis down to the level of the root apical meristem. In intact roots, the root cap forms a complete cone around the root tip. (b) Radial cross-section through the maturation zone. (c) Longitudinal cross-section through the root tip.

Development of root tissues

The developing root consists of three relatively distinct zones (Fig. 1.5). At the apex there is a **zone of cell division**. This includes the apical meristem and extends for up to 1 mm into the developing root tissues. Behind the zone of cell division there is a fairly distinct **zone of cell expansion** in which division has virtually ceased. Finally, behind this, there is a **maturation zone** in which the final stages of cell differentiation take place. The exact positions of these three zones can vary between species, growth conditions and tissue layers.

The development of the epidermis consists of relatively uniform cell division and then cell expansion near to the root tip. Following this, a subset of cells differentiate into **root hairs** in the maturation zone.

Beneath the epidermis, one or more layers of **parenchyma** make up the cortex (parenchyma is a general term used to describe tissues consisting of thin-walled, morphologically undifferentiated cells). Inside the cortex, there is a single layer of endodermis. Endodermal cells lay down a band of cork in the cell walls that are oriented at right angles to the surface of the root. This band, called the **Casparian strip**, blocks the passage of water through the walls of endodermal cells. As a consequence, water and minerals absorbed from the soil must pass through the cytoplasm of the endodermal cells before reaching the vascular tissue. This allows the endodermal cells to regulate the passage of solutes into the vascular system and up into the shoot. In the majority of angiosperms the outermost layer of the cortex, called the **exo-dermis**, also develops a Casparian strip in older non-absorbing parts of the root, where it apparently acts to reduce water loss.

At the centre of the root tip, the root meristem contributes cells to the procambium from which a core of vascular tissue develops, called the **stele**. In most plants, a single bundle of **xylem** forms in the middle of the stele. The bundle of xylem consists of a central region of large-vesselled **metaxylem**, from which arms, or 'poles', of small-vesselled **protoxylem** project. Bundles of **phloem** (which begin to differentiate before the xylem) lie between the protoxylem poles. In each bundle, small-vesselled **protophloem** develops near the edge of the stele, and large-vesselled **metaphloem** develops close to the central xylem. Surrounding both the xylem and the phloem, the stele contains a single layer of parenchyma cells called the **pericycle**.

Root branching

Lateral roots develop in the region behind the band of root hair differentiation. Lateral roots are initiated by a renewal of cell division in the pericycle, normally involving pericycle cells close to one of the protoxylem poles. The developing lateral root grows out through the cortex and epidermis of the parent root to reach the soil. In many plants, cells in the endodermis of the parent root also divide so that lateral roots remain sheathed by endodermis until they break through the parental epidermis.

In some species, roots also produce adventitious shoot buds. These form from shoot apical meristems that usually develop within the pericycle or the cortex of the root. In some trees, shoot buds may form from tissues in the outer bark of a root following secondary thickening.

Primary development of shoots

Shoot apical meristem

Shoots consist of leaf-bearing **nodes** separated by leafless **internodes** (Fig. 1.4). The shoot apical meristem (SAM) is at the very tip of the shoot and is usually considered to consist of the cells above the youngest **leaf primordium**. Seen from above, most SAMs are circular in outline. Seen from the side, they may be convex, flat or concave.

The SAM has both radial and vertical structure (Fig. 1.6). Considering radial structure, the meristem typically possesses: (i) a **central zone**, which consists of large, slowly dividing cells; and (ii) a **peripheral zone**, in which cells divide more rapidly and are usually smaller. The peripheral zone initiates leaves, axillary buds and the outer layers of the stem. Considering vertical

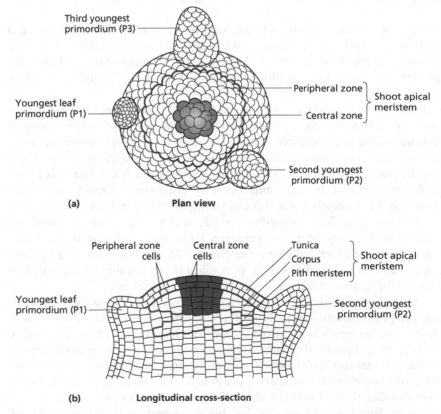

Fig. 1.6 A stylized representation of a dicot shoot tip: (a) plan view and (b) longitudinal cross-section. In most dicots, the shoot apical meristem and young leaf primordia are enclosed by developing older leaves to form the apical bud.

structure, at the meristem surface there are one or more distinct cell layers, collectively called the **tunica**, whereas cells in deeper regions of the meristem are more randomly positioned, forming the **corpus**. The radial and vertical structures of the meristem overlap such that the peripheral and central zones each contain cells from both the tunica and corpus.

In dicots, there is a further group of rapidly dividing cells beneath the central zone that represent the **pith meristem**. These give rise only to the pith of the stem. In monocots, which rarely have a well-defined, central pith (see the description of internode development below), the central tissues of the stem are produced by a large meristematic zone below the apical meristem. This is called the **primary thickening meristem** and can extend both across the diameter of the shoot and for some way down its sides, producing tissue that makes the monocot stem both longer and thicker.

Nodes

Nodes are the sites of leaves and axillary buds. A glance at almost any shoot will show leaves arranged in a regular pattern called the **phyllotaxy**. In some plants, leaves grow as pairs or in groups at each node, but the most common phyllotaxy is a spiral with a single leaf at each node.

Leaves The leaves of dicots and monocots develop somewhat differently and even within each group there is wide variation. Tobacco provides a good example of the basic pattern of dicot leaf development (Fig. 1.7), and the growth of the maize leaf illustrates leaf development in monocots (Fig. 1.8).

Because the absolute rate of leaf growth varies widely with environmental conditions, it is conventional to describe the time course of leaf development in units called **plastochrons**. A plastochron is the period between the initiation of two successive nodes on the shoot. The first plastochron of leaf development in tobacco produces a swelling called a **foliar buttress** in the peripheral zone of the meristem. Normally, the foliar buttress is referred to simply as P1, indicating that it is the youngest visible leaf primordium. The second youngest leaf primordium is P2, and so on. Over the next two plastochrons, the primordium elongates to become peg-shaped. When the leaf is between three and four plastochrons old, the leaf blade (the **lamina**) appears as two bulges flanking the upper face of the primordium, i.e. the face nearest the centre of the apical meristem (the **adaxial** face).

The tobacco leaf develops due to cell division and cell expansion throughout the primordium. The pattern of development is complex with localized, short-lived increases in the division rate in different parts of the leaf lamina. Overlapping these local variations, however, there is a more general pattern. As the leaf matures, cell divisions cease at the tip of the leaf first, and then in a wave moving down to the base of the petiole. The final stages of cell expansion and differentiation occur after cell division stops in each region.

Under sunny conditions, the leaf blade consists of clearly stratified cell types. There is an **upper (adaxial) epidermis**, a **mesophyll** that consists of an upper layer of **palisade parenchyma** and a lower layer of **spongy**

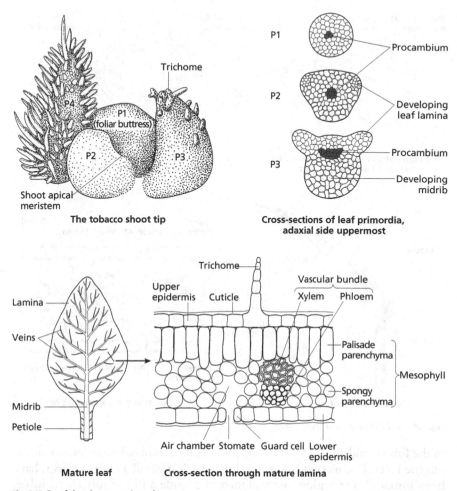

Fig. 1.7 Leaf development in tobacco.

parenchyma (these terms are often simplified to the 'palisade layer' and 'spongy mesophyll', respectively), and then a **lower (abaxial) epidermis.** The epidermis contains **guard cells** around **stomata**, relatively unspecialized **ground tissue** and hairs (**trichomes**, a term used for a variety of epidermal outgrowths including spines, bladders, scales and glands). The mature leaf also has extensive vascular tissue, both in the midrib and as a network of veins in the blade.

The best-studied monocot leaf is that of maize (Fig. 1.8). The maize leaf has two distinct regions. The top of the leaf forms the **blade**, while the base of the leaf wraps around the maize stalk to form the **sheath**. The junction between blade and sheath is clearly marked by two wedges of tissue called **auricles**, and by a flap on the adaxial side called the **ligule.**

The development of the maize leaf begins with the appearance of a primordium that encircles the apical meristem in an overlapping, 'key-ring' shape. A strand of procambium running into the primordium marks the site

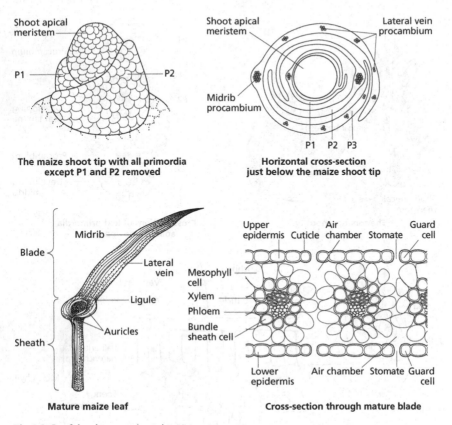

Shoot apical meristem

P1

P2

The maize shoot tip with all primordia except P1 and P2 removed

Shoot apical meristem

Lateral vein procambium

Midrib procambium

P1   P2   P3

Horizontal cross-section just below the maize shoot tip

Blade

Midrib

Lateral vein

Ligule

Auricles

Sheath

Mature maize leaf

Upper epidermis   Cuticle   Air chamber   Stomate   Guard cell

Mesophyll cell

Xylem

Phloem

Bundle sheath cell

Lower epidermis   Air chamber   Stomate   Guard cell

Cross-section through mature blade

Fig. 1.8 Leaf development in maize.

of the future midrib. By the time the primordium is three to four plastochrons old the **lateral veins** (which run up the maize leaf parallel to the midrib) have been initiated, the region that will form the ligule and the auricles is visible, and the cell divisions that will create the ligule are beginning.

Early in maize leaf development, cell divisions take place throughout the primordium. Later, following the same pattern as the tobacco leaf, cell divisions end in a basipetal wave that begins at the leaf tip and moves progressively towards the leaf base. Although cell divisions have ended in the tip of the leaf by the time the primordium is about 3 cm long, the tip region continues to grow extensively by cell expansion before the final stages of cell differentiation occur. Similar waves of expansion and differentiation follow the cessation of cell divisions down the leaf. As a result, the development of the blade finishes before the development of the sheath. The pattern is so marked that there is a period in the leaf's development when the blade is fully mature but cells are still dividing at the base of the sheath.

The epidermis of the maize leaf contains regular arrays of stomata and ground cells, and may also produce leaf hairs. The pattern of cells in the epidermis is related to the position of veins beneath, for example rows of stomata run up the leaf blade but are never directly over a leaf vein. On the inside, the maize leaf has a specialized internal structure called **Kranz** anatomy, which relates to its C4 metabolism. In Kranz anatomy, leaf meso-phyll cells develop in more or less concentric layers around the vascular bundles (Kranz is German for 'wreath'). Furthermore, the innermost one or two layers of mesophyll form a **bundle sheath** in which cells are tightly packed and have thick walls and abundant chloroplasts, mitochondria and other organelles. (The combination of C4 metabolism and Kranz anatomy is an adaptation that reduces the carbon lost by photorespiration.)

The vast range of shapes and sizes of **simple leaves** in both monocots and dicots develops by alterations in the patterns of cell division and expansion in the later stages of leaf development. Some leaves also undergo localized cell death which creates perforations. A more fundamental change, however, occurs in the development of **compound leaves**. In most plants with com-pound leaves, leaf primordia branch to produce secondary outgrowths along their edges. Depending on the complexity of the compound leaf, these outgrowths may branch further or they may develop into leaflets, each following a developmental pattern similar to that of a simple leaf.

Axillary buds and adventitious roots Lateral buds occur in the axils of some or all leaves. They develop from an **axillary meristem**, i.e. a new SAM that is produced in the axil of a leaf, typically simultaneously with leaf development. The time course of initiation and subsequent development of the axillary meristem varies greatly between species. The meristem initiates several leaves to form a bud which subsequently grows into a lateral shoot. Development may be arrested at any point in this series, and commonly stops at the bud stage. A major factor controlling the activity of the bud is the health and proximity to the primary shoot apex: an effect called **apical dominance**. Buds may remain dormant until apical dominance is broken by increased distance from the shoot tip, or loss of the shoot tip.

Nodes are also common sites for the initiation of **adventitious roots**. The apical meristem of the adventitious root may develop at the same time as the other structures of the node, or it may form later in development from older tissues that become meristematic.

Internodes

Radial pattern A cross-section through the stem at an internode shows an outer layer of epidermis that contains stomata and may contain secretory cells and trichomes. Deeper in the stem there is vascular tissue, parenchyma and a variety of supporting tissues. In dicots, the vascular tissue forms either as a continuous cylinder or as a ring of **vascular bundles**. It separates the outer **cortex** of the stem from the central **pith** (Fig. 1.9). In most monocots, vascular bundles develop in a scattered pattern throughout the stem so there is no distinction between an outer cortex and an inner pith. Vascular bundles in the stem contain protoxylem and protophloem at their edges, and metaxylem and metaphloem in the middle. Overall, the bundles are oriented with the phloem nearest to the epidermis and the xylem closest to the centre of the stem. In most dicots, but not in monocots, a residue of meristematic cells remains between the xylem and phloem to form the **vascular cambium**.

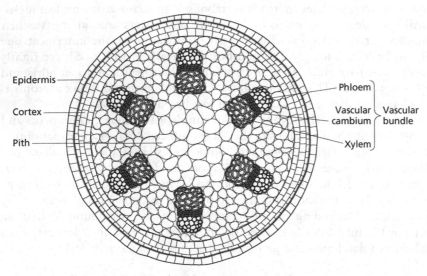

Epidermis

Cortex

Pith

Phloem

Vascular cambium ⎱ Vascular
              ⎰ bundle

Xylem

(a)   **The dicot internode after primary development (radial cross-section)**

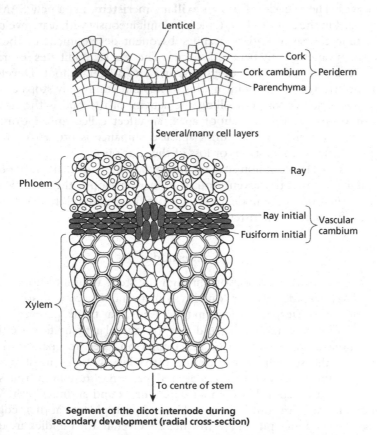

Lenticel

Cork ⎱
Cork cambium ⎬ Periderm
Parenchyma ⎰

Several/many cell layers

Phloem

Ray

Ray initial ⎱ Vascular
Fusiform initial ⎰ cambium

Xylem

To centre of stem

(b)   **Segment of the dicot internode during
        secondary development (radial cross-section)**

Fig. 1.9 (a) Primary and (b) secondary development of the dicot internode.

The discrete vascular bundles seen in stem cross-sections are part of a network of vascular tissue created as vascular bundles branch and fuse. The development of this network is coordinated with the emergence of leaves and buds. As a leaf primordium and its axillary bud appear, the strands of procambium that will eventually connect them to the vascular system develop in the internode below. Procambial strands also form in other regions of the stem and bypass the leaf and bud. Each of these strands connects to existing vascular tissue in the stem, and so vascular tissue in internodes is initiated as a continuous wave moving towards the apex of the shoot.

Elongation Shoots that do not elongate significantly are called **short shoots** whilst those that do elongate are called **long shoots**. Elongation occurs somewhat independently in individual internodes. For each internode, elongation normally begins after several younger internodes have formed and it results from a combination of transverse cell division and longitudinal cell expansion. Cell divisions normally cease at the base of the internode first and at the top last (although some grasses show the opposite pattern with tissue at the base of the internode remaining meristematic even after the upper section is fully mature). Generally, cell expansion overlaps the period of cell division in an internode and then continues for a time after divisions have ended.

Secondary vegetative development

Many dicots and some monocots undergo secondary vegetative development, known descriptively as secondary thickening. In a few monocots, secondary development can be extensive, although sometimes difficult to distinguish from primary development. Palm trees, for example, increase in girth by the prolonged action of an enlarged primary thickening meristem. In most monocots, however, secondary development is very limited. Groups of parenchyma cells in the stem and/or root become meristematic to form clusters of vascular cambium that produce additional vascular bundles in a scattered pattern.

In dicots, secondary development is often extensive and typically involves the action of two cylindrical meristems: an inner cylinder of **vascular cambium** which produces secondary vascular tissue, and an outer cylinder of **cork cambium** which produces a protective layer of cork that replaces the epidermis (Fig. 1.9). The vascular cambium contains vertically elongated cells (called **fusiform initials**) that produce the xylem and phloem, and horizontally elongated **ray initials** that produce rays of parenchyma. The rays penetrate the xylem and phloem and transport water and nutrients between them.

The cork cambium produces cork on its outside and, normally, a single layer of parenchyma on its inside. The three layers, cork, cork cambium and parenchyma, are collectively called the **periderm**. Breaks called **lenticels** form in the outer layer of cork allowing aeration of the tissues beneath.

The ability of dicots but not monocots to undergo extensive secondary thickening helps to explain the different growth habits displayed by the two groups. In dicots, the root system often develops around the thickened

primary root (derived from the embryonic radicle) from which many lateral roots emerge. In monocots, however, the primary root cannot grow wide enough to support the whole root system. Instead it is normally short-lived and the root system develops from adventitious roots formed on the shoot of the embryo and/or seedling. Above ground, the lack of secondary growth makes it mechanically difficult for monocots to support branches. Monocots with aerial branches usually have either very slender stems, supporting structures such as prop roots, or they are climbers.

Transition from the juvenile to the adult phase of the shoot

Shoot development after germination consists of a **juvenile phase** followed by an **adult phase**. The juvenile phase ranges in extent from a few days or weeks in annuals, to many years in some trees. As the name implies, under natural conditions only the adult phase of the shoot can produce flowers. However, the transition from juvenile to adult development can involve many other changes (Fig. 1.10). The most visible difference is often in

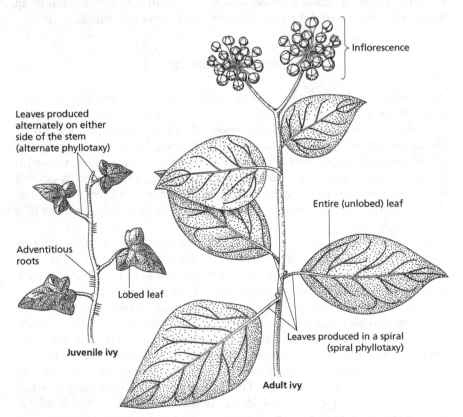

Fig. 1.10 The distinctive juvenile and adult phases of ivy (*Hedera helix*). The juvenile shoot has lobed leaves, alternate phyllotaxy, produces adventitious roots, and grows as a climbing vine. The adult shoot has entire leaves, spiral phyllotaxy, does not form adventitious roots, and grows as a semi-erect shrub.

the shape and/or arrangement of the leaves on the shoot. Lower (i.e. juvenile) sections of the shoot typically have smaller and simpler leaves than higher, adult sections, a condition known as **heteroblasty**. These changes extend to the SAM, which is usually larger in the adult phase than in the juvenile phase.

A further common difference between the juvenile and adult phases of shoot growth is that juvenile stems produce adventitious roots more readily than do adult stems. Other changes associated with the transition follow less general patterns. They include such things as differences in growth habit, changes in the chemical composition of leaves, differences in photosynthetic ability, and altered susceptibilities to diseases and herbivores.

Long-lived, woody plants often show more dramatic differences between juvenile and adult shoots than do herbaceous plants. However, even ephemeral species show some distinctions between the juvenile and the adult phase, implying that the phenomenon is universal.

Floral development

The most common form of the flower, and the one from which most other forms evolved, is radially symmetric and contains four whorls of floral organs (Fig. 1.11). There is an outer whorl of **sepals** (the **calyx**) followed by a whorl of **petals** (the **corolla**). Together the sepals and petals form the **perianth**. Inside the whorl of petals there is a whorl of **stamens** (the **androecium**)

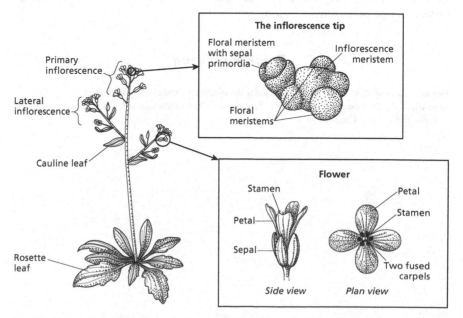

Fig. 1.11 Flowering in *Arabidopsis*. During vegetative development *Arabidopsis* forms a rosette at ground level. On the transition to reproductive development, the plant bolts to produce inflorescences on long stems.

and then a central whorl containing one or more **carpels** (the **gynaecium**). In some primitive flowers, the four organ types form a continuous spiral and the number of floral organs may vary widely. In most species, however, the whorls are separate from each other and the number of floral organs in a particular whorl is essentially invariant.

The huge range of different flowers is mostly the result of modifications to this basic pattern. Floral organs may fuse, either within or between whorls. Flowers may lack either stamens or carpels, and so be unisexual. Parts of the perianth may be missing or much reduced and many flowers have evolved bilateral rather than radial symmetry.

Depending on the species, the transition from vegetative to reproductive development may involve the SAM, axillary meristems, or both. Following the transition to reproductive development, a meristem may produce a single flower or an **inflorescence**. In either case, the switch to reproductive development is usually accompanied by a dramatic increase in the rate of cell divisions over the whole meristem, particularly in the previously slowly dividing cells of the central zone. This causes the meristem to grow bigger and often to change its shape, becoming more highly domed, or flatter, or even both, one after the other.

In plants that produce flowers singly, the meristem is now a **floral meristem** and it initiates primordia, usually from the edge and progressing inwards, that develop into the floral organs. In species that develop flowers in inflorescences, the meristem is now an **inflorescence meristem**. The primordia it produces may develop directly into flowers, i.e. the primordia are themselves floral meristems. More often, however, the primordia grow into leaves or bracts, in the axils of which either floral meristems or lateral inflorescence meristems develop.

Further reading

For an excellent descriptive account of plant development, the reader is referred to:
Steeves, T.A. & Sussex, I.M. (1989) *Patterns in Plant Development*, 2nd edn. Cambridge University Press, Cambridge.

CHAPTER 2

Characteristics of plant development

Chapter 1 described the range of organs produced in the course of the flowering plant life cycle, outlining the core of what is achieved by the mechanisms that govern development. If these mechanisms are to be understood fully, however, other characteristics of plant development must also be considered. Of particular significance at the cellular level are the cell wall and **plasmodesmata**—threads of cytoplasm that breach the wall and connect neighbouring cells. More generally, the continuous nature of plant development, its responsiveness to the environment, and the extraordinary regenerative abilities of plants must be taken into account.

This chapter considers the cellular and larger scale characteristics of plant development and, in the light of these characteristics, outlines a theoretical framework for the discussion of developmental mechanisms.

Plant cells

Cell walls

A wall surrounds all somatic plant cells. For most cells, this wall consists of a single layer, called the **primary cell wall** that is created by the division plate (also called the **phragmoplast**) during cell division and is maintained by deposition of new wall material as the cell expands (Fig. 2.1). On the inside of the primary wall, some cells also produce a thicker, more rigid layer called the **secondary cell wall**. Both layers of the wall consist of bundles of cellulose **microfibrils** embedded in a complex **matrix** of other compounds.

The physical properties of the primary and secondary walls are governed in part by the orientation of microfibrils. In the primary wall, microfibrils are normally laid down at right angles to the long axis of an elongating cell. This creates a 'hoop-like' arrangement of microfibrils which favours further elongation along the long axis. As elongation progresses, however, existing

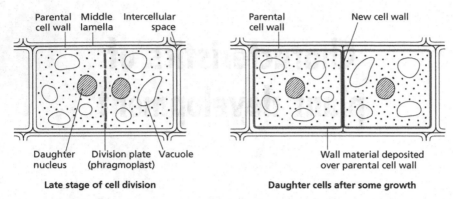

Fig. 2.1 Formation and maintenance of the primary cell wall.

microfibrils are pulled into orientations closer to the cell's long axis while new microfibrils continue to form perpendicular to the axis. As a consequence, the primary wall of elongated cells often has a criss-cross pattern of recent over older microfibrils. The development of microfibrils in secondary cell walls is also highly ordered. The microfibrils typically develop in parallel arrays and the orientation of these arrays can vary between different sublayers of the secondary wall. This creates a laminated structure that increases the wall's strength and rigidity.

The bulk of the matrix in primary walls consists of polysaccharides, mostly **pectins** and **hemicelluloses**. Pectins also make up most of the material in the **middle lamella**, the layer that separates the primary walls of neighbouring cells (Fig. 2.1). In addition to polysaccharides, the matrix of primary walls contains a variety of structural and enzymatic proteins.

In secondary cell walls, the phenolic polymer **lignin** often replaces pectins in the matrix, a process that dramatically alters the wall's properties. Lignification makes cell walls much more rigid and, unless the secondary wall develops in an annular or spiral pattern, it prevents further cell expansion. Lignification also changes the wall from hydrophilic to hydrophobic and impedes the supply of water and solutes to the cell. Extensive lignification is usually associated with cell death.

The existence of the cell wall has profound implications for the control of development. Although differential cell expansion can lead to sections of a plant cell sliding past neighbouring cells, the cell wall essentially precludes cell migration (which is central in the control of animal development). Since primary development occurs around the apical meristems, the growth of shoots and roots is linear with the youngest tissues at the apices. Therefore, during primary development, the age of a tissue is directly related to its distance from the apical meristem. This can make it difficult to separate temporal from spatial aspects of development.

The interconnected cell walls of the plant form the **apoplast** which acts as both a barrier and a channel for intercellular signals. The upper size limit for passage through the apoplast varies between different cell types but it is normally between 10 kDa and 50 kDa, hence blocking the movement of large nucleic acids and proteins. In addition, most cell walls have a net negative charge so they impede the passage of positively charged molecules. However, for small, neutral or negatively charged molecules, the apoplast provides a water-filled channel that allows intercellular movement unimpeded by cell membranes or protoplasm.

The apoplastic channel can be blocked by extensive lignification, or by the deposition of waxy substances such as **cutin** in the cuticle or **suberin** in Casparian strips in the root endodermis. With the exception of tissues isolated by such blocks, the apoplast is continuous with the lumen of xylem vessels. Hence molecules secreted into the cell wall may be transported via the xylem, providing one route for long-range developmental signals (see also the phloem, below). Transport in the xylem occurs in a root-to-shoot direction in the transpiration stream taking water absorbed by the roots to the leaves.

Cell walls could theoretically act as both a repository and a source of developmental signals. Molecules immobilized in the wall matrix could help to fix cell fates. For example, such molecules could interact directly with the abutting cell, or they could affect the activity of other growth regulators arriving through the apoplast. Alternatively, enzymes secreted into the wall could release developmentally active chemicals. Since daughter cells inherit sections of the maternal wall (Fig. 2.1), a cell could influence the fate of its descendants by altering the wall's composition. This is discussed further in Chapter 4 (see case study 4.1).

Plasmodesmata

Most plasmodesmata (singular: plasmodesma) form during cell division. Pores in the division plate turn into channels through the wall, which develop into plasmodesmata connecting the two daughter cells. Each plasmodesma is lined by a membrane continuous with the plasma membranes of the connected cells. In addition, plasmodesmata have an inner, tubular membrane, often called the **desmotubule**, which connects to the endoplasmic reticulum of the two cells (Fig. 2.2). Although most plasmodesmata are initially linear, in maturing tissue they often develop branches and/or fuse with one another. These modifications involve enzymes that digest new channels in the cell wall, into which outgrowths of the original plasmodesmatal membrane and desmotubule project. Plasmodesmata can also arise *de novo* in existing cell walls. This creates cytoplasmic connections between tissue layers that would otherwise be isolated from each other, such as the layers of the tunica in the shoot apical meristem (SAM). It often also occurs during tissue fusion, for example, following the fusion of floral organs in the development of tubular flowers such as *Petunia* and *Antirrhinum*.

The connected protoplasm of cells throughout the plant forms the **symplast** and provides an alternative to the apoplast for the transmission of intercellular signals. The symplast includes the cytoplasm of phloem sieve elements and so provides another route besides the xylem for long-range

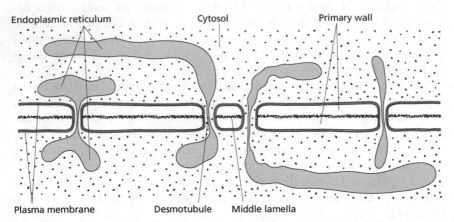

Fig. 2.2 Cross-section through primary, unbranched plasmodesmata. Note that although the endoplasmic reticulum forms a complex, three-dimensional network in each cell, in cross-section only small segments are seen. (Adapted from Lucas *et al.* 1993.)

signalling. The primary physiological function of the phloem is the transport of sucrose and amino acids from mature leaves, where they are made, to 'sinks' such as young leaves and the roots. Hence molecules in the phloem may move in either direction relative to the main root-to-shoot axis of the plant. (It is worth noting that the xylem and phloem do not provide the exclusive means for long-distance transport. The hormone auxin, for example, is synthesized in young leaves and transported all the way to the root apices by parenchyma associated with the vascular tissue. This process involves auxin movement through both the symplast and the apoplast.)

Symplastic communication includes both passive diffusion and active transport of molecules through plasmodesmata. Passive diffusion does not appear to occur through the centre of the desmotubule since its membranes press together and block the central lumen. However, molecules can diffuse through the space between the desmotubule and the outer plasmodesmatal membrane. The size limit for passive diffusion through this space can be tested by injecting different sized dye molecules, or by transforming plants to produce fluorescent proteins of different sizes. Such experiments show that molecules of up to 50 kDa can diffuse through plasmodesmata in developing leaves of tobacco, whereas in mature leaves the limit is 1 kDa or less. This change is associated with the transition from simple to branched plasmodesmata and may profoundly affect the nature of symplastic signalling as the leaf matures.

Dramatic changes in the size limit for passive diffusion also occur during the development of some cell types. In developing guard cells, **callose** deposition entirely blocks plasmodesmata. This isolates guard cells both osmotically and electrically from neighbouring cells, allowing the rapid turgor changes involved in opening and closing stomata. At the other extreme, the cytoplasmic channels through **sieve plates** in the phloem arise from greatly widened plasmodesmata. Sieve plate channels may be several microns across. In contrast, plasmodesmata have a typical diameter of 30–60 nm.

In addition to the passive movement of small molecules, some large nucleic acids and proteins are actively transported across plasmodesmata. Both the prevalence and the mechanism of active transport are currently unknown. Much of the research into the process has focused on viral **movement proteins** required for the systemic spread of some plant viruses. These proteins can transport themselves and viral nucleic acids across plasmodesmata. This transport appears to require targeting motifs on the movement protein and probably involves interactions with the cytoskeleton of the host cell. For cells in mature leaves, the injection of viral movement proteins appears to increase the passive diffusion limit of plasmodesmata from around 1 kDa to about 20 kDa. However, it is not clear if this increase is a necessary part of the active transport process.

The initial effects of injecting viral movement proteins into a cell occur within a few minutes. This implies that the movement proteins are acting in concert with an endogenous system that transports proteins and nucleic acids between cells. Endogenous active transport across plasmodesmata has now been demonstrated, revealing, for example, the transport of proteins and mRNAs from companion cells to sieve elements in the phloem, and, most likely, the transport of proteins between layers of the SAM.

Active transport into sieve elements allows the long-range movement of mRNAs and proteins via the phloem, although the developmental significance of this is not yet clear. There is also strong evidence for short-range active transport of transcription factors (proteins that regulate gene transcription) between layers of the SAM. For example, the gene encoding the KNOTTED1 (KN1) transcription factor, which is required for SAM function (see case study 5.2), is only expressed in subsurface layers of the maize SAM. However, KN1 protein can be detected in all layers of the SAM, including the outermost, suggesting that KN1 moves from underlying layers into the surface layer through plasmodesmata. Supporting this theory, KN1 injected into leaf mesophyll cells can move through plasmodesmata into neighbouring cells.

It has been proposed that the presence of plasmodesmata makes it inappropriate to view plants as truly cellular and that plant development is best considered at a supracellular level. Support for this view comes from evidence that regions of a plant may form **symplastic domains**. For example, in the apex of the potato shoot there is a partial symplastic barrier blocking the movement of injected dye molecules between the apical meristem and emerging leaf primordia. One way to interpret this observation is to view the meristem and the primordia as separate symplastic domains within which development occurs at a supracellular level. However, there is ample evidence that neighbouring cells can act autonomously despite plasmodesmatal connections. So, although symplastic domains may have an important role in development, it is necessary to consider development and developmental mechanisms at the cellular level.

Larger patterns

Continuous development

Plant development continues throughout the life cycle. Embryogenesis establishes the shoot–root axis, the radial pattern of tissues and the first apical meristems. However, almost all plant organs, i.e. roots, stems, leaves and flowers, form after germination in a process that is largely reiterative. Apical meristems repeat the same, or similar, developmental patterns to produce an extending root or an extending series of nodes and internodes. To a large extent, lateral meristems also reiterate the pattern of development around the apical meristem. The combination of continuous, reiterative development reaches its extreme in the widespread occurrence of vegetative reproduction.

Open-ended patterns of development are called **indeterminate** whereas patterns restricted in time or space are called **determinate**. For example, development at the shoot apical meristem (SAM) is often indeterminate but the development of a leaf primordium is usually determinate. The mechanisms controlling plant development must coordinate indeterminate and determinate development. The development of the plant as a whole may follow either pattern. For example, perennial species normally have indeterminate growth while many annuals have determinate growth.

Developmental patterns may also switch from indeterminate to determinate, or vice versa. A vegetative SAM may become determinate if it becomes a floral meristem. If it becomes an inflorescence meristem it may be determinate or indeterminate, depending on the species. In contrast, otherwise determinate plant organs may give rise to indeterminate adventitious root or shoot buds.

Plastic development

Because development is continuous, most plant organs develop exposed to the vagaries of the environment. The development of each organ, and of the plant as a whole, can therefore be adjusted according to the prevailing environmental conditions—an important skill for a sedentary organism. For example, the common garden weed couch grass (*Agropyrum repens*) spreads by underground shoots consisting of long internodes and nodes that bear scale leaves and adventitious roots. On reaching the soil surface, such shoots produces short internodes and rootless nodes that bear true leaves (Fig. 2.3). This is a response to the change from dark to light conditions, i.e. to the change in light intensity. Other environmental cues that regulate plant development include the direction and spectral quality of incident light; the duration of the photoperiod; gravity; mechanical stress; temperature; the concentration of carbon dioxide in the atmosphere; humidity; and localized variations in soil water and nutrient availability. Developmental modifications also occur in response to other organisms, for example mycorrhizal fungi and nodulating bacteria.

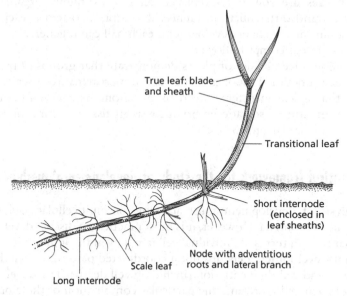

Fig. 2.3 Plastic, reiterative development of couch grass (*Agropyrum repens*). Couch grass spreads by underground shoots (rhizomes) consisting of long internodes and nodes that bear scale leaves and adventitious roots. In response to light at the soil surface, the shoot switches to forming short internodes and rootless nodes bearing true leaves.

Environmental responsiveness can be achieved either through modifications to a single developmental pattern or by switches between apparently distinct developmental programmes. This distinction is somewhat subjective. For example, **tropic growth**, in which part of the plant bends in response to a directional signal, usually results from differential cell expansion in a developing organ. This is clearly a minor modification to development. In contrast, the differences between an expanded leaf produced by a tree in spring and a bud-scale produced on the same branch in summer or autumn represent extensive changes in both cell proliferation and differentiation. Even though bud-scales evolved from leaves, these changes are probably best considered as a switch between separate developmental programmes.

Regeneration

It is likely that all living plant cells are **totipotent**. Under the correct conditions they would be able to grow from a single cell to a complete plant. In the natural environment, of course, plant regeneration is much less extreme and usually represents either damage repair or vegetative reproduction.

Regeneration after damage normally occurs in two stages. Firstly, parenchyma cells near to a damaged site proliferate to produce disorganized **callus** tissue. Within the callus, cells then organize and differentiate in ways that depend upon the position of the wound, for example producing new shoot or root apical meristems, and/or new vascular strands. However, the production of callus is not essential for regeneration. In developing organs, missing tissues are sometimes replaced directly by highly organized cell proliferation and differentiation. Hence, if a floral meristem is excised and bisected at an early stage of development, each half can regenerate in culture to form an almost complete flower.

The regenerative abilities of plants demonstrate that groups of plant cells can produce functional shoot and root apical meristems from scratch. This suggests that apical meristems are quasi-autonomous and self-organizing. The phenomenon of somatic embryogenesis in tissue culture implies that embryos have similar properties.

Theoretical framework for the study of developmental mechanisms

Any analysis of development must account for multicellular, cellular and subcellular phenomena. However, it is most practical to discuss developmental mechanisms in terms of individual cell fate. Given the immobility of plant cells, plant development depends upon coordinated patterns of cell division, cell growth and cell differentiation (including cell death). The **developmental fate** of each cell represents the particular combination of these processes displayed by the cell.

At the biochemical level, these cellular behaviours are determined by the pattern of gene expression in the cell. This is very often regulated at the level of gene transcription. Consequently, in many instances cell fate depends on the production or activity of transcription factors. Furthermore, individual transcription factors often control several or many of the genes that specify particular cellular behaviours and therefore act, either individually or in combination, as master regulators of cell fate. The study of development has therefore been focused on the mechanisms controlling the activity of these master regulators.

There are a number of experimental approaches that can be used to investigate the control of master regulators and their effects on cell fate and many of the terms used in developmental biology have experimental definitions that arose from the application of such approaches to the study of animal development. Consequently, commonly used terms are not always appropriate in the analysis of plant development. For example, in animal embryology, a cell is said to be **determined** if its developmental fate remains constant when the cell is excised from the embryo and is grown in tissue culture. In plants, however, such experiments are difficult to interpret. In tissue culture almost any living plant cell can be induced to form callus from which entire plants may regenerate. From a strict experimental perspective, therefore, determination rarely occurs in living plant cells. Yet this does not mean that the fate of plant cells remains unrestricted during normal development. It simply proves that plants have profound regenerative abilities.

Given such difficulties in transferring experimentally defined terms from animal development to plant development, this book uses a framework for categorizing developmental mechanisms based on purely theoretical concepts.

Cell-intrinsic and cell-extrinsic information

Theoretically, the mechanisms controlling cell fate fall into two classes. Such mechanisms may depend upon information originating within the cell, i.e. **cell-intrinsic information**; alternatively, they may rely upon information originating from outside the cell, i.e. **cell-extrinsic information**. Somewhat arbitrarily, this book considers information present in the cell wall as extrinsic.

Intrinsic information could arise during a cell's individual history, for example in the form of factors relating to the cell's age. Alternatively, a cell could inherit intrinsic information from its ancestors, or be subject to stochastic cell-intrinsic changes. Information carried within the cell could be carried by any aspect of intracellular composition, for example the organization of the cytoskeleton, the nutrient status of the cell or epigenetic changes to the cell's DNA (such as DNA methylation).

Cell-extrinsic information may either arise within the plant or come from the environment. Cell-extrinsic information originating within the plant could involve short-range signals between neighbouring cells, long-range signals that arrive from more distant parts of the plant, or even the generation of physical forces within a tissue. Extrinsic information originating in the environment includes gravity, temperature, the intensity and spectral quality of light, and so on.

This theoretical classification provides a useful framework around which to consider the experimental evidence. However, it must be stressed that mechanisms based upon cell-intrinsic and cell-extrinsic information are frequently interdependent. For example, signals from the environment are transduced into signals within the plant and these eventually affect the internal state of individual cells. Furthermore, the internal state of a cell affects its production of, and responses to, external signals.

Conclusions

Plants are sessile organisms unable to avoid either the vagaries of the environment or physical harm. They have evolved developmental patterns that are continuous, responsive to external information, and flexible in the event of damage. As will become clear in later chapters, these attributes are founded on the extensive exchange of information between plant cells, both locally and at long range. Signalling between cells occurs via the apoplast, which is composed of interconnected cell walls, and the symplast, which is composed of the interconnected cytoplasms of plant cells.

Further reading

Cell walls

Brett, C. & Waldron, K. (1990) *Physiology and Biochemistry of Plant Cell Walls.* Unwin Hyman, London.

Cosgrove, D.J. (1997) Assembly and enlargement of the primary cell wall in plants. *Annual Review of Cell and Developmental Biology* **13**, 171–201.

Plasmodesmata

Lucas, W., Ding, B. & Van Der Schoot, C. (1993) Plasmodesmata and the supracellular nature of plants. *New Phytologist* **125**, 435–476.
Oparka, K.J., Roberts, A.G., Boevink, P. *et al.* (1999) Simple, but not branched, plasmodesmata allow the nonspecific trafficking of proteins in developing tobacco leaves. *Cell* **97**, 743–754.
Pickard, B.G. & Beachy, R.N. (1999) Intercellular connections are developmentally controlled to help move molecules through the plant. *Cell* **98**, 5–8.
Ruiz-Medrano, R., Xoconostle-Cázares, B. & Lucas, W.J. (1999) Phloem long-distance transport of CmNACP mRNA: implications for supracellular regulation in plants. *Development* **126**, 4405–4419.

Symplastic transcription factor movement in the SAM

Lucas, W.J., Bouché-Pillon, S., Jackson, D.P. *et al.* (1995) Selective trafficking of KNOTTED1 homeodomain protein and its mRNA through plasmodesmata. *Science* **270**, 1980–1983.
Perbal, M-C., Haughn, G., Saedler, H. & Schwarz-Sommer, Z. (1996) Non-cell-autonomous function of the *Antirrhinum* floral homeotic proteins DEFICIENS and GLOBOSA is exerted by their polar cell-to-cell trafficking. *Development* **122**, 3433–3441.
Sessions, A., Yanofsky, M.F. & Weigel, D. (2000) Cell–cell signaling and movement by the floral transcription factors LEAFY and APETALA1. *Science* **289**, 779–781.

Larger patterns of development

Walbot, V. (1996) Sources and consequences of phenotypic and genotypic plasticity in flowering plants. *Trends in Plant Science* **1** (1), 27–32.

CHAPTER 3

Cell-intrinsic information

Since plant cells cannot migrate, each cell in the plant must adopt a fate appropriate to its immediate position. The next four chapters consider how this is achieved. As discussed in Chapter 2, cell fate could depend on cell-intrinsic information, cell-extrinsic information, or both. Chapters 4, 5 and 6 consider the role of cell-extrinsic positional information in the specification of cell fate. This chapter discusses how information intrinsic to the cell might be involved.

Cell-intrinsic information can be classified according to the means by which it is acquired. The first part of this chapter considers the role of information inherited from the mother cell. This information connects the fate of a cell to its lineage. The second part discusses the role of age-related information. The chapter does not discuss the role of stochastic processes occurring within the cell. However, stochastic changes in either cell-intrinsic or cell-extrinsic information can be important in establishing the pattern of some plant structures and this topic is discussed in Chapter 6.

Lineage

In animals there is good evidence for lineage-dependent mechanisms determining cell fate. This is reflected in the fact that only the cells of the early embryo are totipotent, being able to give rise to all cell types. As embryogenesis proceeds, the range of fates open to cells progressively decreases so that they are restricted first to subsets of fates and then to a particular fate by their lineage. Such lineage restrictions are brought about by a variety of mechanisms. For example, fate-determining factors may be asymmetrically segregated into the daughter cells during cell division; or fates may be imposed by extrinsic positional information and then passed on to daughters.

In contrast, the role of lineage in plants is far from certain. One reason for this is that the immobility of plant cells makes it very difficult to distinguish between cell lineage and cell position, and hence between the influence of lineage on cell fate and that of extrinsic positional information. This section begins by considering the relationship between cell lineage and cell position. It then discusses whether lineage influences the fate of plant cells.

Relationship between cell lineage and cell position

A long tradition of analysing slices of plant tissue has created a fairly comprehensive picture of where cells divide, expand and differentiate. However, individual slides can give only a static picture of development and they are an impractical way of tracing detailed lineage patterns. Instead, the relationship between cell lineage and cell position is usually studied using a technique called **clonal analysis** (Fig. 3.1). The basic approach is to label a cell at some early stage in development with a heritable, visible marker. This allows the distribution of the descendants of the marked cell to be determined at later stages in development. The most common way to mark cells is through the use of randomly generated somatic mutations in genes involved in pigment production, such as those affecting chlorophyll or anthocyanin synthesis. Somatic mutations may occur spontaneously; for example, due to the excision of a transposable element. Alternatively, they may be induced by mutagens such as X-rays.

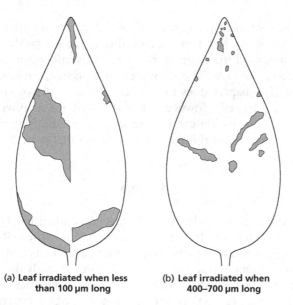

(a) Leaf irradiated when less
than 100 µm long

(b) Leaf irradiated when
400–700 µm long

Fig. 3.1 Clonal analysis of tobacco leaf development. Clones of cells with aberrant chlorophyll synthesis were generated by exposure to X-rays. The figure shows marked clones in the palisade layer (pink). Clones induced early in development (a) are larger than those induced later (b) because more cell divisions occur before the leaf matures. Clones near the tip of the leaf in (b) are smaller than those near the base because the leaf matures in a basipetal wave (Chapter 1). (Adapted from Poethig & Sussex, 1985.)

Interpreted in the light of detailed anatomical information, clonal analysis has shown that there can be a very strong relationship between cell lineage and cell position. Examples of this are described below.

Fate of lineages derived from the shoot apical meristem

The relationship between the position of a cell in the shoot apical meristem, and the fate of its descendants in the shoot, is revealed by plants in which the apical meristem is a mosaic of marked and unmarked cells (Fig. 3.2). A plant such as this, made up of two or more genetically distinct populations of cells, is called a **chimera**. The two most simple types of chimera are **sectorial chimeras**, in which different circumferential sectors of the meristem consist of differently marked cells (like slices of a cake), and **periclinal chimeras**, in which vertical layers of the meristem contain differently marked cells (like the sugar and chocolate layers of a Smartie or M&M).

In sectorial chimeras, marked circumferential sectors of the apical meristem initiate bands of tissue that run approximately vertically down the shoot (Fig. 3.2). This suggests that the shoot is divisible into circumferential lineages, each of which is derived from a small number of initial cells at the top of the meristem. Presumably, these initials represent all or a subset of the central zone (see Chapter 1). Sectorial chimeras can exist for many

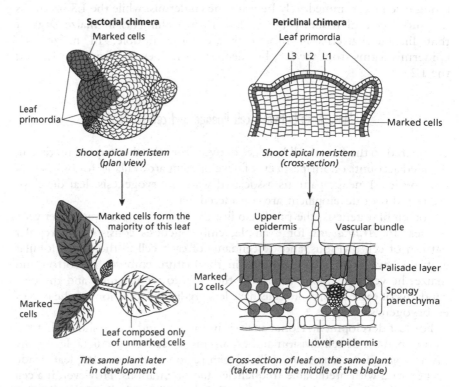

Fig. 3.2 Sectorial and periclinal chimeras of tobacco.

nodes but they are not usually permanent, suggesting that function as an initial cell depends on a cell's position within the meristem. When a marked initial is displaced by an unmarked cell, then within that layer of the meristem (see below) the marked sector disappears.

Sectorial chimeras show that the circumferential position of an initial in the apical meristem does not restrict the fates of the initial's descendants (Fig. 3.2). The position of the marked band in a sectorial chimera is independent of the pattern of leaf initiation. Hence, on successive nodes of a single chimeric shoot, the marked band may encompass a whole leaf, part of a leaf, or be confined entirely to the stem. The cells in the band are not restricted by their lineage to producing only leaves or only stem tissue.

In contrast, periclinal chimeras reveal that lineages derived from different layers of the apical meristem normally generate characteristic tissues in the shoot (Fig. 3.2). As described in Chapter 1, the shoot apical meristem consists of two relatively distinct regions—an outer, layered tunica and an inner cell mass called the corpus. The majority of dicots have two discrete layers of cells in the tunica, whereas most monocots have only a single tunica layer. Starting at the surface, the cell lineages represented by the layers of the apical meristem are called the L1, the L2 and the L3. In dicots, the two tunica layers are the L1 and the L2, and the corpus is the L3. In monocots, the tunica is the L1 and the corpus is the L2.

Periclinal chimeras in dicots show that the shoot epidermis develops from the L1, and that for significant sections of the shoot, the L2 forms a single cell layer immediately beneath the epidermis, while the L3 produces the more central tissues (Fig. 3.2). Periclinal chimeras in maize suggest that lineage patterns in monocots are more flexible. In maize, the epidermis forms from the L1, but deeper tissues develop from the L1 and the L2.

More links between cell lineage and cell position

Compared to the fate of cell lineages derived from the shoot apical meristem, detailed accounts of other aspects of development are available for only a very few species. Lineage patterns associated with embryogenesis, leaf development and root development are considered here.

For embryogenesis, the predictability of lineage patterns varies between species. In *Arabidopsis*, for example, embryogenesis consists of a regular pattern of cell divisions, and descendants of each cell in the early globular embryo have predictable positions in the mature embryo. In cotton and maize, however, division patterns in the embryo are irregular and the descendants of early embryonic cells have less predictable positions at the end of embryogenesis.

For leaf development, clonal analysis in tobacco and maize shows that the overall pattern of cell division and expansion is relatively predictable. At any given stage of development, cells in each region of the growing leaf divide and expand with predictable frequencies and orientations. However, if a cell is marked early in leaf development, the exact dimensions of the clone it produces are unpredictable. Therefore, lineage predicts the approximate but not the exact positions of leaf cells. As an exception to this rule, in several species stomatal complexes (consisting of guard cells and surrounding epidermis) and multicellular leaf hairs (trichomes) represent clonal groups. The stomatal complex or the trichome arises from a mother cell by a characteristic pattern of cell divisions. However, cell lineage does not appear to determine which epidermal cells will become the trichome or stomatal mother cells.

Lineage patterns in the root have also been studied in a few species. In *Arabidopsis*, distinct sets of initials in the apical meristem produce specific tissues in the root, creating a strong link between the lineage and position of root cells (see Fig. 1.5). The *Arabidopsis* root apical meristem consists of initial cells surrounding a quiescent centre. On the basal face of the meristem (that pointing *away* from the root tip), a central group of cells gives rise to vascular tissue, and a peripheral ring of cells initiates the endodermis and cortex. On the apical face of the meristem (that pointing *towards* the root tip), a central group of cells initiates the central region of the root cap, and a peripheral ring of cells gives rise to the lateral sections of the root cap and to the epidermis.

It is unlikely that the root apical meristems of all species will display such regular lineage patterns. *Arabidopsis* has a type of root apical meristem called a **closed** meristem, a pattern shared by other species, such as maize. However, many species have an **open** meristem in which distinct sets of initials are not observed. The lineage patterns in open root apical meristems are unclear, although regular cell division patterns have been observed in open meristems above the disorganized initial region.

Lineage case studies

The predictable lineage patterns associated with many aspects of development are consistent with lineage-dependent mechanisms influencing cell fate. The remainder of this section considers investigations into whether such mechanisms exist.

Case study 3.1 discusses laser ablation, the technique of using a fine laser beam to kill either individual cells or very small groups of cells. Laser ablation is useful for testing the influence of lineage on cell fate. After the treatment, dead cells collapse and neighbouring cells are forced into the gap. If the fate of the invading cells is determined primarily by their lineage, then the cells' behaviour should be unchanged by the experiment. However, if cell fate depends mainly upon extrinsic positional information, the invading cells will adopt fates appropriate to their new positions. Laser ablation experiments reveal that the invading neighbours act according to their new positions rather than their lineage. By implication, the position of a cell can override its lineage in determining the cell's fate.

The results of laser ablation must be interpreted with caution because they represent the plant's response to damage (albeit, minimal damage), rather than its normal developmental pattern. As an alternative to laser ablation, the relation between lineage and cell fate has also been studied by observing the effects of unusual division patterns on cell behaviour. Case study 3.2 discusses spontaneous deviations from typical division patterns, and case study 3.3 analyses the effects of mutations that disrupt division patterns. Both approaches have the advantage of being non-invasive and provide results uncontaminated by the plant's damage responses. Overall, the investigations support the conclusions of laser ablation experiments: cell fate depends primarily on extrinsic positional information rather than on cell lineage.

It remains likely, however, that lineage influences cell fate to some extent. To adopt a particular fate, a cell must express the set of genes appropriate to that fate. Hence cells in different regions of the plant display different gene expression patterns. It is probable that these differences are inherited through cell divisions. This would give cells a lineage-based identity. The observed absence of lineage-dependent cell fate restrictions implies that inherited information is cross-checked against external positional cues and can be overridden when discrepancies occur.

Case study 3.1: Laser ablation of cells in the *Arabidopsis* root tip

During laser ablation, a very fine laser is used to kill cells in a developing organ. Cells killed by the laser collapse and pressure within the plant pushes neighbouring cells into the space created. As a result, the technique can be used to study the relative importance of lineage and position to cell fate. Do the invading cells behave according to their lineage, or do they assume fates appropriate to their new location? Laser ablation in the *Arabidopsis* root tip suggests that position rather than lineage ultimately determines cell fate.

In the core of the *Arabidopsis* root, the apical meristem produces root cap cells on its outer face and procambium cells (vascular precursors) on its inner face. Between these two meristematic zones lies the quiescent centre in which divisions are rare or absent. The influence of lineage on cell fate has been tested by ablating the quiescent centre (Fig. 3.3). The dead quiescent centre cells collapse and cells from the procambium take their place. The effects of this change on the fates of the invading cells can be observed by using transgenic plants carrying artificially constructed marker genes that are active specifically in vascular cells, or in quiescent centre cells, or in cells of the central region of the root cap (the **columella**).

In plants carrying the vascular marker gene, the marker is initially active in the invading procambium cells but becomes inactive in the days following the experiment. In contrast, in plants carrying either the quiescent centre or the columella marker gene, the marker is initially inactive in the invading cells but becomes active, in appropriate relative positions, in the days following the experiment (Fig. 3.3). By implication, the displaced procambium cells switch from vascular to either quiescent centre or columella fates, indicating that their lineage does not restrict them to producing vascular tissue. As a consequence, a functional root apex is regenerated.

Further laser ablation experiments reveal that similar flexibility in cell fate exists along the radial axis of the root. *Arabidopsis* roots have a very simple

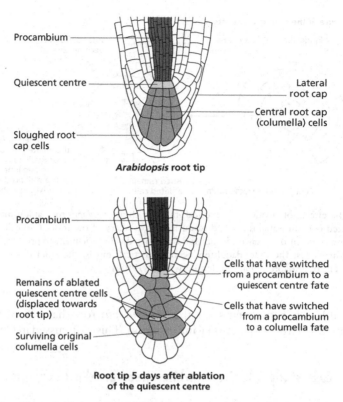

Arabidopsis **root tip**

**Root tip 5 days after ablation
of the quiescent centre**

Fig. 3.3 The effects of ablation of the quiescent centre in the *Arabidopsis* root tip. The ablated quiescent centre cells are crushed and displaced towards the root tip. As a consequence, procambium cells come to occupy more distal (tipwards) sites and switch fates to produce a new quiescent centre and central root cap (columella).

radial pattern of tissues (see Fig. 1.5). There is an outer layer of epidermis, a single layer of cortex, a single layer of endodermis, a single pericycle layer and a central core of vascular tissue. Both the cortex and the endodermis trace their lineage back to a single ring of cells in the apical meristem called the **cortical initials**. In the meristem, this ring of initials lies immediately outside a ring of smaller, pericycle cells. If a cortical initial is ablated, two of the smaller pericycle cells may be pushed outwards into the dead cell's place (Fig. 3.4). The invading pericycle cells then divide **periclinally** (in a plane parallel with the root surface). The inner daughter cells continue to act as part of the pericycle, but the outer daughter cells become new cortical initials. Lineage restrictions, therefore, do not appear to apply to cell fates along the radial axis of the root.

Overall, the results of laser ablation imply that cell fate depends on cell position rather than cell lineage. By implication, cell fate depends primarily on cell-extrinsic positional information. This assumes, however, that damage responses caused by laser ablation do not overcome normal lineage constraints on cell fate.

Radial structure of the root apical meristem

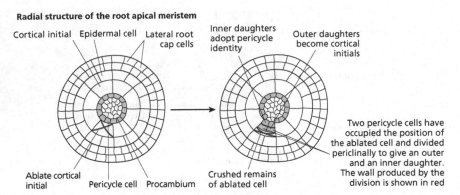

Fig. 3.4 The effects of ablation of a cortical initial in the *Arabidopsis* root apical meristem. A single ablated cortical initial is crushed towards the periphery of the root. Two cells from the pericycle are forced into the site of the ablated cell. They divide periclinally to produce outer and inner daughter cells: the outer daughters become cortical initials, the inner daughters adopt pericycle identity. (Adapted from van den Berg *et al.*, 1995.)

Laser ablation experiments on the *Arabidopsis* root have been used to investigate the nature of positional information. This is discussed in Chapter 4.

Case study 3.2: Green–white–green periclinal chimeras

The assumption that damage responses induced by laser ablation do not remove lineage constraints can be questioned. The behaviour of the invading cells after laser ablation represents localized regeneration of the root. Yet it is known that regeneration on a larger scale, through the production of callus, can produce entire plants, obviously overcoming any lineage constraints present in the cells from which the callus formed. By itself, therefore, laser ablation cannot prove that lineage does not restrict cell fate. However, in conjunction with the results of the non-destructive experiments discussed in this case study and the next, this conclusion is inescapable.

The least disruptive way to study the influence of lineage on cell fate is to observe the effects of spontaneous deviations from typical lineage patterns. This case study considers how such deviations affect the development of 'green–white–green' (GWG) periclinal chimeras. In these chimeras, the L1 and L3 of the shoot apical meristem are wild type but cells of the L2 are unable to synthesize chlorophyll. GWG chimeras have been investigated in several dicot species to uncover the influence of an L1, L2 or L3 lineage on cell fate. The results are unequivocal. Although deviations from the normal lineage pattern are rare, the cells affected always develop in accordance with their new position rather than with their L1, L2 or L3 lineage.

For example, GWG chimeras show that although the L2-derived cells are largely confined to a single cell layer immediately beneath the epidermis (the **subepidermal** layer), L2 tissue frequently invades the central L3 core of the shoot. L2 invasions are clearly visible in the leaves of GWG plants (Fig. 3.5).

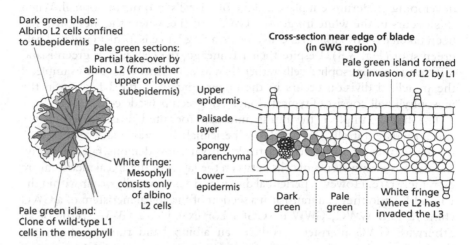

Fig. 3.5 A GWG *Pelargonium* leaf. The cross-section shows the take-over of the mesophyll by the albino L2 to produce the white fringe. The pale green island consists of green L1-derived cells that have been placed in the palisade layer by a periclinal division. (Adapted from Stewart, 1978.)

In a typical dicot, the L1 forms the upper and lower epidermis of the leaf blade throughout most of the blade, the L2 is confined to the subepidermal layers, and the L3 forms the central core of the mesophyll. Except for guard cells, epidermal cells do not produce chlorophyll. Therefore, the colour of the leaves on GWG chimeras depends upon the relative contributions of the L2 and the L3 to the underlying tissues. When both L2 and L3 cells are present, the blade is green because green L3 cells are visible beneath the albino L2 cells. However, if L2 cells invade deeper tissues and take over from the L3, the leaf blade is white.

GWG chimeras show that although L2 cells remain in the subepidermal layers in most of the leaf, periclinal divisions near the edge of the blade place L2 cells into deeper tissues where they adopt fates typical of L3 cells. There is a compensatory reduction in cell divisions in the L3 so that the number of cell layers in the leaf remains constant. The result is a variegated blade with a green centre, showing the presence of L3; and a white fringe, demonstrating take-over of the mesophyll by L2-derived cells. This pattern is often seen in variegated garden plants, for example in many variegated varieties of ivy, privet, *Pelargonium*, etc. The size of the L2 fringe can vary extensively from leaf to leaf, and even between different parts of the same leaf. Similar L2 invasions occur in other parts of GWG shoots leading, for example, to albino sections of the stem. However, these invasions never affect the final anatomy of the shoot. Lineage does not restrict L2 cells to a subepidermal fate. Instead, invading L2 cells adopt fates appropriate to their new positions, for example producing vascular tissue in the central mesophyll.

In dicots, the L1 very rarely extends beyond the epidermis into deeper parts of the shoot. Yet GWG chimeras show that when such invasions occur, L1 cells also develop according to their position rather than their lineage. During the development of the leaf blade, rare periclinal divisions in the developing epidermis can place a clone of L1 cells into the mesophyll. When this occurs in the white fringe of a GWG leaf (i.e. where the mesophyll has been taken over by albino L2 cells) the invading L1 cells form an isolated pale green spot (Fig. 3.5). Despite their L1 lineage, the cells in this green island differentiate as mesophyll cells rather than as epidermal cells. For example, if the periclinal division occurs in the upper epidermis, the L1 clone in the mesophyll will consist of typical elongated, green palisade cells.

The final example is the rarest: unusually oriented divisions within the apical meristem itself. The stability of periclinal chimeras, a fact witnessed by the endurance of variegated horticultural varieties, demonstrates that divisions transferring clones of cells between the layers of the apical meristem are extremely rare. However, periclinal divisions do occasionally occur within the L1 and L2 and they can transform a section of the apical meristem of a GWG chimera into WWG, GWW or GGG. For example, a GWW sector of an otherwise GWG meristem produces an albino band running down the shoot. In contrast, those parts of a leaf that are within a GGG sector of the shoot are solid green rather than variegated. In all cases, the anatomy of the shoot is unaffected, implying that the invading cells act according to their new position. Therefore, lineage patterns created by the layered structure and regular planes of cell division of the shoot apical meristem do not correspond to lineage constraints on cell fate.

Coordination of growth between layers of the shoot

As discussed above, when L2 cells from the subepidermal layer invade the core of the mesophyll, there is a compensatory reduction in division in the L3 such that the morphology of the leaf is normal. Similarly, when L1 cells invade the subepidermal layer, growth in the L2 is correspondingly reduced. This suggests that signalling within tissue layers coordinates total cell proliferation in each layer. Similarly, relative growth rates are highly coordinated between layers, as observed in polyploid/diploid chimeras.

In all eukaryotes, cell size is roughly proportional to ploidy; for example, diploid ($2n$) cells are about four times smaller by volume than octoploid ($8n$) cells of the same cell type. In the late 1930s, Satina, Blakeslee and Avery generated a variety of ploidy chimeras in the thorn apple (*Datura*) (Fig. 3.6). These included $8n$ (octoploid)-$2n$ (diploid)-$2n$ (L1–L2–L3), $2n$-$8n$-$2n$ and $2n$-$2n$-$8n$ periclinal chimeras. It was through these experiments that the division of the shoot into the L1, L2 and L3 lineages was discovered.

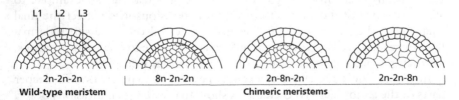

<table>
<tr><td>L1 L2 L3</td><td></td><td></td><td></td></tr>
<tr><td>2n-2n-2n</td><td>8n-2n-2n</td><td>2n-8n-2n</td><td>2n-2n-8n</td></tr>
<tr><td>**Wild-type meristem**</td><td colspan="3">**Chimeric meristems**</td></tr>
</table>

Fig. 3.6 Octoploid/diploid chimeras of thorn apple (*Datura*). (Adapted from Satina *et al.*, 1940.)

Despite the gross differences in cell sizes in polyploid/diploid chimeras, the overall morphology of the shoot is normal. For example, in an $8n$-$2n$-$2n$ chimera, L1 cells are about four times larger than L2 and L3 cells, but the surface area of the epidermis is approximately normal due to a corresponding reduction in the number of L1 cells. This suggests that the growth of the epidermis is regulated by signals from deeper tissue. Similarly, in $2n$-$8n$-$2n$ chimeras and $2n$-$2n$-$8n$ chimeras, the growth of tissues derived from the L2 and L3, respectively, is normal despite their large cell size due to compensatory decreases in cell number.

Case study 3.3: Mutations affecting division patterns

The flexibility of cell fate demonstrated by spontaneous deviations from typical division patterns is striking. For a wider test of the influence of lineage, this case study considers mutations that cause far more dramatic disruptions to normal lineage patterns. Maize plants mutant at the *TANGLED1* gene (see genetic nomenclature, p. 42) have disrupted longitudinal and transverse cell divisions throughout development. Similarly, mutations in the *FASS* gene of *Arabidopsis* randomize the orientation of most cell divisions. The development of mutant plants shows that an approximately normal pattern of tissues and organs can accompany even grossly abnormal cell lineages, making it highly unlikely that lineage restricts cells to specific cell fates in developing regions of the plant.

tangled1 mutants of maize

Mutations in the *TANGLED1* (*TAN1*) gene of maize disrupt the orientation of the new cell wall formed during cell division, and consequently change overall division patterns. *tan1* plants are smaller than wild-type plants and have dramatically altered cell shapes. Because of the changes in cell division patterns, it can be assumed that the mutants have unusual lineage patterns. Despite this, the overall shape of *tan1* plants is normal, suggesting that predictable lineage patterns are not needed for plant organs to develop in their usual arrangement, and into their usual shapes. Therefore, cells must adopt fates appropriate to their position in the plant rather than their lineage.

The effects of *tan1* mutations have been studied in most detail for the leaf blade (Fig. 3.7). In wild-type maize, the developing blade displays a regular pattern of cell divisions, with most cells dividing either longitudinally or transversely. Since cells are aligned in files with their long axis parallel to the long axis of the blade, longitudinal divisions increase the number of cells across the blade, while transverse divisions lead to more cells along the length of the blade. The long, thin shape of a maize leaf blade reflects the fact that transverse divisions outnumber longitudinal divisions at all stages of the blade's development.

tan1 mutations disrupt the orientation of the division plate. In addition to longitudinal and transverse divisions, many cells in the developing blade

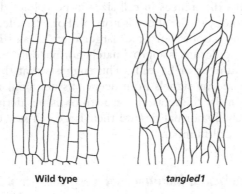

Wild type   *tangled1*

Fig. 3.7 The epidermis of wild-type and *tangled1* maize leaf primordia. Despite the irregular arrangement of cells, *tangled1* primordia develop into correctly shaped leaves. (Adapted from Cleary & Smith, 1998.)

undergo aberrant divisions in which the division plate is crooked or curved. The result is a change both in cell shape and in the relationship between cell lineage and cell position. Yet the overall shape of the leaf blade is indistinguishable from that of the leaf blades of wild-type plants. The mutations affect the development of roots, stems and floral organs in a similar way. Again, the gross appearance of mutant plants is normal. Therefore, plant morphology is at least partly independent of the usual links between a cell's position and its lineage. The development of *tan1* plants implies that lineage mechanisms play at most a minor role in determining organ arrangement and shape.

tan1 mutations do disrupt the fine structure of plant organs. For example, the upper epidermal surface of an adult maize leaf consists of roughly parallel ranks of 'ground' cells punctuated at regular intervals by trichomes and stomata. In contrast, the epidermis of mutant blades contains a patchwork of irregularly shaped cells and has a disordered distribution of trichomes and stomata. The pattern of internal tissues is also irregular—the name 'tangled' refers to the chaotic arrangement of leaf veins in mutant plants. It is likely that these disruptions reflect difficulties in organizing irregularly shaped cells rather than the effects of altered lineage patterns.

Coordinated control of cell division and cell expansion

Apart from their significance for the study of lineage mechanisms, *tan1* mutants also demonstrate how closely the control of cell expansion and cell division are linked. Since the orientation of cell divisions is disrupted in *tan1* plants, the patterns of cell expansion must compensate in order to develop organs with a wild-type shape.

The ability of changes in cell expansion to compensate for changes in cell division is also observed in other contexts. For example, when cell division in a wheat embryo is blocked by gamma radiation, leaf initiation can occur through cell expansion alone (Fig. 3.8). Prior to germination, the majority of wheat embryos possess three true-leaf primordia. During the germination

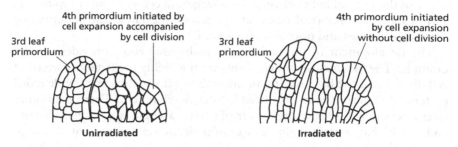

Fig. 3.8 The effect of gamma radiation on leaf initiation in wheat embryos. (Adapted from Foard, 1971.)

of an unirradiated seed, a fourth primordium emerges accompanied by cell division and cell expansion in the apical meristem. Amazingly, when wheat grains are subjected to gamma radiation sufficient to prevent cell division but not to kill the wheat embryo, a fourth primordium still arises in the correct location but this time solely as a result of cell expansion. In the absence of cell division, leaf development does not proceed past this stage.

In the reverse case, the large size of polyploid cells in polyploid/diploid periclinal chimeras is compensated for by a reduction in cell division (see case study 3.2). Other examples of the coordination of cell growth and division are discussed with respect to the control of leaf size in case study 5.3.

fass mutations in *Arabidopsis*

In contrast to the relatively normal appearance of *tan1* maize plants, mutations in the *FASS* gene of *Arabidopsis* generate plants that are extremely short and unusually thick; the gene is named after the German word for 'barrel' (Fig. 3.9). In *fass* plants, cells divide and expand in seemingly random orientations at almost all stages of development, presumably destroying

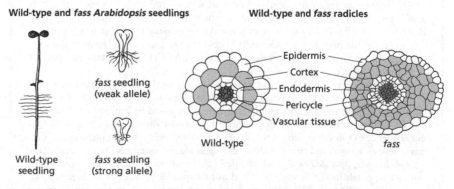

Fig. 3.9 The effects of *fass* mutations of *Arabidopsis*. *fass* mutants have randomly oriented cell division and cell growth making them far shorter and broader than the wild type. Despite this, mutants develop the correct longitudinal and radial patterns of tissues and organs. (Adapted from Torrez-Ruiz & Jürgens, 1994; Scheres *et al.*, 1995.)

many of the normal links between cell lineage and cell position. Despite this, and despite the abnormal shape of the plants, the radial and longitudinal patterns of tissues and organs are preserved.

The development of the radicle (the embryonic root) provides a good example. The *Arabidopsis* radicle forms from a highly predictable pattern of cell divisions in the basal section of the embryo. It has the same simple radial pattern of tissues as roots produced in later development. There is an outer layer of epidermal cells; a single layer of cortex; a single layer of endodermis; and a stele that contains a single layer of pericycle cells and a central core of vascular tissue.

In *fass* mutants, most cell divisions in the embryo are oriented abnormally. Mutant embryos have a greatly thickened radicle containing several extra layers of cells, yet the radicle possesses the normal radial pattern of tissues. There is an outer layer of epidermis, several layers of cortical cells, a single layer of endodermis, a single pericycle layer, and an enlarged bundle of vascular tissue.

A similar phenomenon occurs in the development of all other parts of *fass* plants. Mutants have thick, distorted roots, stems, leaves and flowers but these all develop in an approximately normal pattern. Furthermore, their tissues lie in the correct positions relative to one another. The disruption to the orientations of both cell division and cell growth must destroy many of the normal links between cell lineage and cell position. Therefore, lineage mechanisms are unlikely to be required to generate the fundamental radial and longitudinal arrangement of tissues and organs.

Box 3.1: Genetic nomenclature

There is no universal system of genetic nomenclature. Geneticists studying different species have different conventions on how to write the names of wild-type and mutant alleles of a gene, or to describe the protein encoded by a gene. For simplicity, this book uses the *Arabidopsis* conventions for all genetic studies. Take the *LEAFY* gene as an example.

The name of the wild-type allele of a gene is written in uppercase italics—*LEAFY* (*LFY*). Mutant alleles are indicated by lowercase italics—*leafy* (*lfy*). Lower case italics are also used to describe the mutations that define mutant alleles—*lfy* mutations; and to describe plants that display the mutant phenotype—*lfy* mutants or *lfy* plants.

A gene and the protein it encodes usually share the same name. The name of the wild-type protein is written in uppercase plain text—LEAFY (LFY). Mutant proteins are indicated by lowercase plain text—leafy (lfy).

It is worth pointing out that most genes are named for their loss-of-function phenotype. This means that the name of a gene often suggests the opposite of the gene's function. On *leafy* mutants, for example, flowers are replaced by leafy shoots. Therefore the *LEAFY* gene is required for flower development (described in case study 9.3). One consequence of this convention is that genes defined by similar mutant phenotypes often have similar names and are therefore distinguished by numbers. For example, the *apetala1* (*ap1*), *apetala2* (*ap2*) and *apetala3* (*ap3*) mutants are all characterized by the absence of petals in the flower (described in case study 5.4) and define three separate genes: *AP1*, *AP2* and *AP3*. Mutant alleles of the same gene are distinguished by a hyphenated number, for example *ap2-2* and *ap2-4* are both mutant alleles of *AP2*.

Relationship between age and position

Because plant cells are immobile and cell division occurs in localized regions, the position of both cells and organs is related to their age. For example, the average age of cells in the shoot or root rises with increasing distance from the apex. Similarly, cell divisions cease at the apex of a developing leaf earlier than at the base (Chapter 1); hence the oldest cells in a fully developed leaf are those at the leaf tip.

The links between cell age and cell position make it difficult to distinguish between age-dependent and position-dependent mechanisms in the determination of cell fate. For example, an age- or a position-dependent mechanism could direct leaf cells to the appropriate fate for their position along the proximodistal (tip-to-base) axis. The problem of distinguishing age- from position-dependent mechanisms is also illustrated by the difficulty of interpreting mutations that cause plant structures to develop at unusual sites. Mutations that vary the position of cell fate aquisition are called **homeotic**, those that vary the timing of cell fate aquisition are called **heterochronic**. In animals, homeotic and heterochronic mutations are usually easy to separate because development is normally restricted to particular stages in the animal's life cycle. When a fruit fly develops with legs in the place of its antennae, this is clearly a homeotic mutation because the timing of development does not alter. In contrast, if petals grow in the place of sepals, the mutation is both homeotic and heterochronic. The petals are unusually placed, but they are also developing at an unusual (earlier) time.

In addition to the possibility that age-dependent mechanisms direct cells to fates appropriate to their position, such mechanisms could also influence the development of the plant as a whole. For example, age-dependent mechanisms could be one of the factors regulating the transitions from juvenile to adult, and from vegetative to reproductive development. Given that the timing of such transitions varies greatly with the growth rate of the plant, occurring later when growth is slow and sooner when growth is fast, it has been suggested that they are influenced by 'developmental' rather than absolute age. The obvious alternative is that the transitions are influenced by the plant's size.

Mechanisms for monitoring either cell or plant age could be based on cell-intrinsic factors, cell-extrinsic factors, or a combination of the two. They are discussed in this chapter because the simplest mechanism to envisage is one based upon a cell-intrinsic timer, perhaps monitoring the steady accumulation or degradation of a particular molecule (the term 'timer' is used rather than the more obvious 'clock' to avoid confusion with the circadian clock, which is discussed in Chapter 7). In the simplest case, a cell-intrinsic timer could affect the individual fate of the cell carrying it. However, a timer could also alter the signals produced by the cell and so affect development on a larger scale. Furthermore, a timer need not be limited to measuring the age of an individual cell; it might also time the period elapsed since a fixed point in development, for example, from germination or from the initiation of an organ.

Currently, the existence of age-dependent mechanisms in plants is largely hypothetical. The single case study contained in this section (case study 3.4) outlines what is probably the clearest evidence for such a mechanism in plant development: the effects of mutations that alter the rate of leaf initiation in *Arabidopsis*.

Case study 3.4: Mutations affecting the rate of leaf initiation in *Arabidopsis*

In *Arabidopsis*, as in many other plants, there are progressive changes in leaf morphology from germination through to flowering. These changes have been used to study the factors affecting shoot maturation, in particular whether changes in vegetative development are driven by the increasing age or the increasing size of the shoot.

Leaf shape and the distribution of trichomes (hairs) on leaf surfaces both change in a predictable manner as the *Arabidopsis* shoot matures. In particular, there is an easily recognizable transition from the production of leaves that have no trichomes on their underside (abaxial side), to the production of leaves in which the abaxial face carries at least some trichomes (Fig. 3.10). The stage of development at which abaxial trichomes appear depends on environmental conditions; however, under any one set of conditions, abaxial trichomes appear at a relatively constant node. For example,

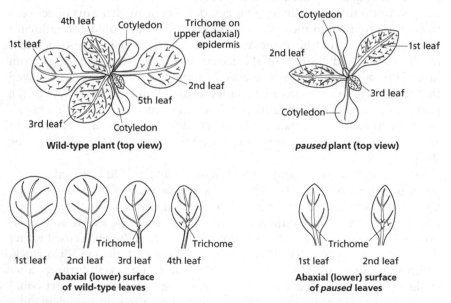

Fig. 3.10 Wild-type and *paused Arabidopsis* at the same age. The first and second leaves of the *paused* seedling were produced at approximately the same age (time since germination) as the third and fourth leaves of the wild-type seedling and have approximately the same shape and trichome distribution as the wild-type third and fourth leaves. (Adapted in part from Telfer *et al.*, 1997.)

when the commonly used *Arabidopsis* variety Landsberg *erecta* is grown under constant light, abaxial trichomes normally appear under the third or fourth true leaf.

The node at which abaxial trichomes appear could depend upon age-based, size-based or position-based information: the third leaf develops when the shoot is at a characteristic age and has reached a characteristic size, but the leaf is also in a characteristic position along the shoot-to-root axis. The effects of the *paused* (*psd*) and *altered meristem programming1* (*amp1*) mutations help to distinguish between these possibilities: both imply that age, not size or position, is the key.

In wild-type plants a pair of true leaves form immediately after germination. In contrast, most *psd* mutants fail to develop true leaf primordia until several days later. This is due to the death of cells in the central zone of the shoot apical meristem of mutant seedlings, and the subsequent delay in development as the meristem regenerates. The cause of cell death in the mutant meristem has yet to be established.

For the purposes of this discussion, *psd* mutants are interesting because the true leaves that eventually grow do not resemble the first true leaves of wild-type plants. Instead, the first leaves produced by most *psd* mutants look like leaves produced later in the development of a wild-type shoot (Fig. 3.10). In particular, the first or second true leaf on a *psd* mutant often has abaxial trichomes. As a result, most *psd* mutants produce leaves with abaxial trichomes at about the same age as wild-type plants. In both wild-type and *psd* plants, leaves with abaxial trichomes appear approximately 7 days after the seeds are planted (with seedlings grown under constant light). The effects of *psd*, therefore, imply that leaf morphology in *Arabidopsis* is influenced by the time since germination, not by the position of the leaf on the shoot or the size of the plant.

The *amp1* mutation supports the conclusion that an age-dependent mechanism influences leaf morphology. The mutation causes leaves to be initiated at about twice the wild-type rate. If leaf morphology depends upon the position of leaves on the shoot, abaxial trichomes should appear earlier in time on *amp1* mutants than on wild-type plants. This does not happen. Instead, *amp1* mutants develop leaves with abaxial trichomes at almost the same time after germination as wild-type plants, even though this means producing about twice as many leaves with bare abaxial surfaces.

The effects of *psd* and *amp1* imply that the *Arabidopsis* shoot possesses a timing mechanism that influences vegetative characteristics according to the plant's age. The mutations also show that the extent of vegetative development, i.e. the size of the shoot, can be uncoupled from this timing mechanism. The effects of *amp1* indicate that the timer does not operate by counting cell divisions, for example, at the apical meristem. In *amp1* plants, cells in the meristem will have divided more often than is normal at the time when abaxial trichomes first appear. If plant cells possess an intrinsic timer, it may be able to pass through successive cell divisions still 'ticking'. There is evidence for such timing mechanisms acting in animal embryogenesis.

The molecular nature of the age-dependent mechanism in the *Arabidopsis* shoot is unknown, however its effects may be related to gibberellin biosynthesis or response. Gibberellins are a class of plant hormones that, among other things, promote the transitions between developmental phases. As will be discussed in Chapter 9, gibberellins promote the transition from embryonic to post-embryonic development, from juvenile to adult development, and from vegetative to reproductive development.

In *Arabidopsis*, the production of abaxial trichomes on leaves is an adult trait. Exogenous application of gibberellins accelerates the production of abaxial trichomes, although never to the first two leaves of a wild-type plant. Furthermore, in mutants deficient in either gibberellin synthesis or response, the production of leaves with abaxial trichomes is delayed. The most severe gibberellin-deficient mutants do not produce abaxial trichomes unless supplied with exogenous gibberellin.

Conclusions

This chapter has considered the role of cell-intrinsic information during development. The overwhelming conclusion is that intrinsic information rarely determines the exact fate of a cell. Morphological, biochemical and genetic studies show progressive cell-intrinsic changes during every aspect of development. However, these changes appear to reflect rather than determine cell fate. It is likely that intrinsic information such as an inherited pattern of gene expression, or an intrinsic timer, can influence cell fate. However, such intrinsic information can be completely overridden by extrinsic signals.

The precise behaviour of cells in developing regions of the plant appears to be determined primarily by extrinsic information. Chapters 4, 5 and 6 now discuss the nature of cell-extrinsic positional information. Chapters 7 and 8 will then consider the plant's responses to environmental cues.

Further reading

General

Lineage

Poethig, R.S. (1987) Clonal analysis of cell lineage patterns in plant development. *American Journal of Botany* 74, 581–594.

Relationship between age and position

Lawson, E.J.R. & Poethig, R.S. (1995) Shoot development in plants: time for a change. *Trends in Genetics* 11, 263–268.
Poethig, R.S. (1990) Phase change and the regulation of shoot morphogenesis in plants. *Science* 250, 923–930.

Case study 3.1: Laser ablation of cells in the *Arabidopsis* root tip

Sabatini, S., Beis, D., Wolkenfelt, H. *et al.* (1999) An auxin-dependent distal organizer of pattern and polarity in the *Arabidopsis* root. *Cell* 99, 463–472.

Van den Berg, C., Willemsen, V., Hage, W., Weisbeek, P. & Scheres, B. (1995) Cell fate in the *Arabidopsis* root meristem determined by directional signalling. *Nature* 378, 62–65.

Case study 3.2: Green–white–green periclinal chimeras

Stewart, R.N. (1978) Ontogeny of the primary body in chimeral forms of higher plants. In: *The Clonal Basis of Development* (ed. S. Subtelny & I.M. Sussex), pp. 131–160. Academic Press, New York.

Stewart, R.N. & Dermen, H. (1975) Flexibility in ontogeny as shown by the contribution of the shoot apical layers to leaves of periclinal chimeras. *American Journal of Botany* 62, 935–947.

Szymkowiak, E.J. & Sussex, I.M. (1996) What chimeras can tell us about plant development. *Annual Review of Plant Physiology and Plant Molecular Biology* 47, 351–376.

Tilney-Bassett, R.A.E. (1986) *Plant Chimeras*. Arnold, London.

Diploid/polyploid chimeras

Satina, S. & Blakeslee, A.F. (1941) Periclinal chimeras in *Datura stramonium* in relation to the development of the leaf and flower. *American Journal of Botany* 28, 862–871.

Satina, S., Blakeslee, A.F. & Avery, A.G. (1940) Demonstration of the three germ layers in the shoot apex of *Datura* by means of induced polyploidy in periclinal chimeras. *American Journal of Botany* 27, 895–905.

Case study 3.3: Mutations affecting division patterns

tangled1

Cleary, A.L. & Smith, L.G. (1998) The *Tangled1* gene is required for spatial control of cytoskeletal arrays associated with cell division during maize leaf development. *Plant Cell* 10, 1875–1888.

Smith, L.G., Hake, S. & Sylvester, A.W. (1996) The *tangled-1* mutation alters cell division orientations throughout maize leaf development without altering leaf shape. *Development* 122, 481–489.

Gamma-irradiated wheat embryos

Foard, D.E. (1971) The initial protrusion of a leaf primordium can form without concurrent periclinal cell divisions. *Canadian Journal of Botany* 49, 1601–1603.

fass

Torres-Ruiz, R.A. & Jürgens, G. (1994) Mutations in the *FASS* gene uncouple pattern formation and morphogenesis in *Arabidopsis* development. *Development* 120, 2967–2978.

Case study 3.4: Mutations affecting the rate of leaf initiation in *Arabidopsis*

Telfer, A., Bollman, K.M. & Poethig, R.S. (1997) Phase change and the regulation of trichome distribution in *Arabidopsis thaliana*. *Development* 124, 645–654.

CHAPTER 4

Primary axis development

The experiments discussed in Chapter 3 indicate that cell fate in developing regions of the plant is highly flexible and determined largely by cell-extrinsic positional information. Chapters 4, 5 and 6 now discuss how cell position is defined in plants.

Detailed models for the definition of cell position have emerged from studies of animal development. Cell position can be defined with respect to a fixed point such as a particular cell, tissue or organ. Alternatively, position can be defined with respect to axes running along and across the organism. These models often rely on gradients of fate-determining substances termed **morphogens**, either emanating from a fixed point or distributed along axes. In such models, cell fate is regulated by the local concentration of the morphogen. Putative examples of morphogen-regulated development have been identified in animals (Chapter 10). However, it is clear from the study of such examples that cell fate is also modulated by extensive, local interactions in the field of cells exposed to the morphogen gradient.

This chapter considers the development of the primary longitudinal and radial axes of the plant. Chapter 5 discusses axes in leaves and flowers. Chapter 6 considers mechanisms that relate cell fate to the proximity of a particular cell, tissue or organ.

Embryonic axes

A plant embryo has two main axes, the **longitudinal axis** running from shoot to root, and the **radial axis** leading from core to circumference. This chapter considers how these axes are established, maintained and elaborated during embryogenesis. It also discusses the relationship between the embryonic axes and the development of the radial and longitudinal axes of post-embryonic shoots and roots.

The axes of the embryo and of post-embryonic shoots and roots are **polar**—cell types at one end of the axes differ from those at the other end. This polarity is also expressed individually by cells along the axes. For example, root hair cells (**trichoblasts**) develop a root hair from their outer face, indicating cell polarity along the radial axis of the root. Furthermore, in *Arabidopsis*, the root hair forms from the end of the trichoblast closest to the root tip (see Fig. 1.5), demonstrating cell polarity along the longitudinal axis.

Cell polarity is not restricted to mature organs; it is also a feature of cells in developing axes. An example of this is asymmetric cell division in which a polarized cell divides to give daughters of unequal sizes, which may adopt different fates. This can be seen in the initial division of the *Arabidopsis* zygote into a large basal cell and a small terminal cell (see Fig. 1.2 and case study 4.2). Hence the formation, maintenance and elaboration of the axes occurs at both the cellular and supracellular levels.

The acquisition of polarity has been studied extensively in the seaweed *Fucus*. The *Fucus* egg is spherical and the zygote acquires longitudinal polarity in response to environmental cues. The polarity of the zygote is then elaborated into the longitudinal axis of the embryo. Case study 4.1 discusses the definition, maintenance and elaboration of longitudinal polarity during *Fucus* embryogenesis. Case studies 4.2 and 4.3 then consider embryogenesis in flowering plants, discussing the development of the *Arabidopsis* longitudinal and radial axes.

Case study 4.1: Longitudinal axis of the *Fucus* embryo

Fucus is a brown alga that grows on rocks high in the intertidal zone. It produces free-floating eggs which are fertilized by motile sperm. After fertilization, the zygote attaches itself to a rock and commences embryogenesis. The *Fucus* egg is spherical and apolar. The zygote acquires longitudinal polarity largely in response to environmental cues. The polarity of the zygote is then converted into the longitudinal axis of the embryo, characterized by an apical embryonic frond, the **thallus**, and a basal embryonic holdfast, the **rhizoid** (Fig. 4.1). Asymmetry in the zygote is established rapidly after fertilization with chloroplasts and maternal mRNA concentrating at the presumptive thallus pole; while mitochondria, actin filaments, microtubules and Golgi activity concentrate at the presumptive rhizoid pole.

The accessibility of the zygote and embryo has allowed extensive experimental investigation into *Fucus* embryogenesis. Research on *Fucus* (and its close relative *Pelvetia*) has shown that the initial polarization of the zygote involves the redistribution of actin filaments (**F-actin**) and the formation of intracellular calcium and pH gradients. Zygote polarity is then converted into asymmetries in the cell wall of the zygote, and between the cell walls of its daughters.

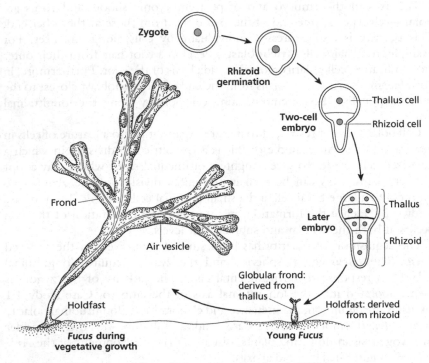

Fig. 4.1 The development of *Fucus*. (Adapted from Kropf *et al.*, 1999.)

Establishing polarity in the zygote

The *Fucus* egg has no cell wall and contains a central nucleus surrounded by a uniformly distributed cytoplasm. The first sign of polarity in the zygote occurs within minutes of fertilization as a patch of F-actin accumulates at the site of sperm entry (Fig. 4.2). In the absence of polarized environmental

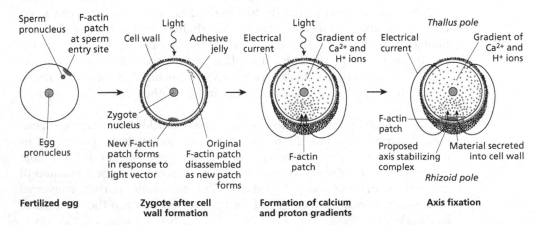

Fig. 4.2 The formation and maintenance of the longitudinal axis in *Fucus*. (Adapted from Kropf *et al.*, 1999.)

cues, this site will become the rhizoid pole of the zygote. In normal circumstances, however, the longitudinal axis is oriented relative to external information.

After fertilization, the zygote wall rapidly forms and the zygote secretes a sticky jelly that attaches it to the substratum. At this point, the longitudinal axis becomes sensitive to environmental cues, particularly to directional light. The illuminated face of the zygote becomes the thallus pole, and the shaded face forms the rhizoid pole. The polarity of the zygote is also influenced by gravity, water currents and temperature gradients.

The mechanism by which the zygote transduces environmental signals into internal polarity is unknown. If the environmentally determined axis is oriented differently from the axis defined by site of sperm entry, the F-actin patch marking the sperm entry site is disassembled and a new F-actin patch accumulates at the new rhizoid pole. Subsequently, an electrical current that flows out of the zygote at the presumptive thallus pole and into the zygote at the presumptive rhizoid pole can be detected (Fig. 4.2). The nature of the current is unknown but it probably involves the movement of calcium and/or hydrogen ions through asymmetrically distributed pumps and channels in the zygote plasma membrane. Intracellular concentration gradients can be measured for both ions, with the highest concentrations in both cases being at the presumptive rhizoid pole. It is likely that the asymmetric distribution of ion pumps and channels is directed by the polarized distribution of F-actin. The central role of the cytoskeleton is a common feature in the establishment and maintenance of cell polarity in a wide range of organisms.

Role of the cell wall in maintaining the longitudinal axis

The polarity of the *Fucus* zygote is initially labile and can be re-oriented for example, by changing the direction of illumination. After each change, a new F-actin patch forms at the new presumptive rhizoid pole, the old patch disappears, and the calcium and pH gradients are re-oriented. However, several hours after fertilization, the longitudinal axis becomes fixed and the positions of the future thallus and rhizoid cannot be altered by external cues. This process is called **axis fixation** and appears to involve interactions between the cytoplasm and the cell wall. Zygote **protoplasts**, i.e. cells from which the wall has been removed, readily acquire polarity, but their polarity remains labile until a new wall has been secreted.

During axis fixation, the patch of actin filaments at the rhizoid pole appears to guide Golgi-derived vesicles, called F-granules, to that region of the zygote, where the vesicles secrete a variety of proteins and polysaccharides into the cell wall. The secretion of adhesives also concentrates at the rhizoid pole. Transmembrane links are believed to form between some of the material secreted into the wall and the actin filaments. It has been suggested that the actin filaments and substances in the cell wall form an 'axis-stabilizing complex' at the rhizoid pole (Fig. 4.2). This hypothesis is supported by experiments involving brefeldin A, a compound which inhibits Golgi activity and hence F-granule secretion. Without the secretion of F-granules into the wall, axis formation occurs as normal but the zygote's polarity remains labile. At present, it is unclear which of the materials secreted into the wall are required for axis fixation. Presumably the complex acts to anchor cytoskeletal elements along the newly formed axis.

Role of the cell wall in elaborating the longitudinal axis

At the time of axis fixation, the zygote is internally polarized but still spherical. In the following hours, the first signs of axis elaboration occur. In a process referred to as germination, the rhizoid pole grows to produce a tubular rhizoid (Fig. 4.1). After the rhizoid emerges, the zygote divides asymmetrically in the plane perpendicular to its longitudinal axis. This produces a large apical cell from which the thallus develops, and a small basal cell from which the rhizoid continues to form. During embryogenesis the thallus develops as a block of tissue, while cells in the rhizoid make predominantly transverse divisions to produce an elongating strand. Probably by coincidence, this pattern is remarkably similar to the development of the embryo proper and suspensor in many angiosperms (see Fig. 1.2).

A simple axis-stabilizing complex would be sufficient to explain how the polarity of the zygote is fixed. However, research on the elaboration of the thallus–rhizoid axis implies that more complex positional information is also secreted into the cell wall.

Experiments on the two-cell embryo show that the walls of both the thallus cell and the rhizoid cell are required to maintain normal development (Fig. 4.3). For example, if a small hole is cut by laser in the wall of the thallus cell, the thallus protoplast can be extruded into the culture medium. The protoplast does not regenerate into a thallus. Instead, it develops a new longitudinal axis and goes on to form a complete embryo. A protoplast derived from the rhizoid cell will behave likewise. In contrast, if a laser is used to remove both the rhizoid cell and the rhizoid cell wall, leaving an isolated but intact thallus cell, thallus development continues as normal. In the same way, an isolated rhizoid cell with an intact wall will continue rhizoid development.

Further laser ablation experiments show that the thallus and rhizoid cell walls can direct as well as simply maintain cell fate (Fig. 4.3). For example, if the rhizoid protoplast is removed from a two-cell embryo but fragments of the rhizoid wall are left attached to the thallus cell, initially divisions in the isolated thallus will proceed as normal. However, when growth of the thallus forces cells against the rhizoid wall fragments, those cells switch to a rhizoid pattern of development. They lose most of their chloroplasts and they change the orientation of their cell divisions to produce an elongating rhizoid. Hence, contact with the rhizoid cell wall can convert a thallus cell to a rhizoid cell.

In the converse experiment, the thallus cell is ablated but the thallus cell wall is left attached to the rhizoid cell. In this case, rhizoid development is initially normal but cells will switch to a thallus fate if they contact the thallus wall. Therefore, both the thallus and rhizoid walls of the two-cell embryo can

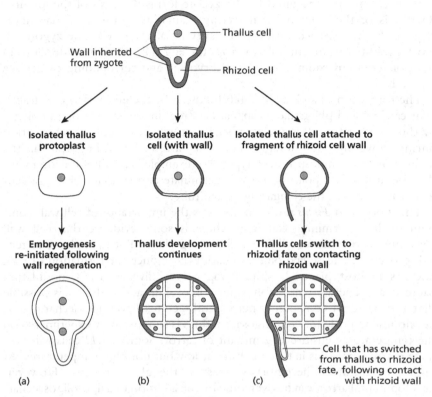

Fig. 4.3 Cell fate can be switched by cell wall contact in *Fucus*. Experiments on thallus cells isolated from two-cell *Fucus* embryos indicate that cell fate is regulated by the cell wall. (a) If the cell wall is removed from an isolated thallus cell to produce a protoplast, the protoplast regenerates a cell wall and then re-initiates embryogenesis. (b) An isolated thallus cell with an intact cell wall continues thallus development. (c) If a fragment of rhizoid cell wall remains attached to an isolated thallus cell, thallus cells that eventually come into contact with the rhizoid cell wall switch to rhizoid fate. (Adapted from Berger *et al.*, 1994.)

cause cells of the opposite type to switch fates. The wall components responsible must be long lived since several days can elapse between cell ablation and fate switching. The wall components must also have the ability to act through the walls of the affected cells.

It is not known when or how the wall components that determine thallus and rhizoid fate arise. Since the outer walls of both the thallus cell and the rhizoid cell in the two-cell embryo are inherited from the zygote, it is possible the fate-determining factors are secreted into the zygote wall before the first cell division.

Extrapolating from *Fucus* to flowering plants

Fucus is a brown alga and not closely related to higher plants, which evolved from the green algae. Therefore, the relevance of research on *Fucus* to the understanding of development in flowering plants is uncertain.

In *Fucus*, polarity acquired by the zygote forms the basis of the longitudinal axis of the embryo. In flowering plants, axis formation may never require the *de novo* acquisition of polarity by a single cell. The zygotes of most angiosperms inherit polarity from the egg cell, which itself develops in the polarized environment of the embryo sac and surrounding ovule (see Fig. 1.1).

The formation of an F-actin patch followed by the generation of intracellular calcium and pH gradients appear key steps in the acquisition of polarity by the *Fucus* zygote. It is unclear if such factors play a role in axis formation during flowering plant development. They are, however, important in the polar growth of angiosperm cell types. For example, a patch of F-actin marks the site at which a pollen grain will germinate and a calcium ion gradient forms at the tip of the elongating pollen tube.

The work with *Fucus* also demonstrates the importance of cell wall components in determining cell fate. There is some evidence that cell wall components may direct cell fate in flowering plants. For example, nitrogen-fixing bacteria release lipo-oligosaccharides to induce the formation of root nodules in host plants. Despite being a very diverse group, nodulating bacteria all generate remarkably similar lipo-oligosaccharides. It is possible that the bacterial compounds mimic endogenous lipo-oligosaccharides. A developmental role for lipo-oligosaccharides is supported by experiments on the temperature-sensitive *ts11* mutant of carrot. Somatic *ts11* cells will produce mature embryos in tissue culture at low but not high temperatures. At high temperatures, the embryos arrest at the globular stage. However, developmental arrest can be overcome by the addition of a lipo-oligosaccharide nodulation factor produced by the nodulating bacterium *Rhizobium*. Furthermore, *ts11* somatic embryos will continue to develop at high temperatures if supplied with a chitinase secreted by wild-type embryos. It is believed that the chitinase releases lipo-oligosaccharides from the plant cell wall.

A further intriguing observation comes from studies in which monoclonal antibodies were raised against angiosperm cell wall polysaccharides. The antibodies were used in *in situ* immunolocalization experiments on a variety of species to determine which cells had specific polysaccharide epitopes in the cell wall. Distinctly different epitope patterns are observed in the walls of different root cells showing that there is cell-type specific variation in cell wall composition. This variation could be involved in the maintenance of cell fate in higher plants; however, as yet there is little direct evidence that this is the case.

Case study 4.2: Longitudinal axis of the *Arabidopsis* embryo

The embryos of flowering plants develop beneath several layers of tissue, making them less accessible to physical or chemical experimentation. Embryos grown in tissue culture circumvent this problem but may not follow the same mechanisms of development as zygotic embryos. Therefore, much of our current understanding of the mechanisms controlling embryogenesis in flowering plants comes from the analysis of embryo mutants.

This case study discusses how the polarity of the *Arabidopsis* egg, and hence of the zygote, is converted into the longitudinal axis of the embryo, and how the axis is maintained and elaborated.

Converting the polarity of the zygote into the longitudinal axis of the embryo

In common with many angiosperms, the egg cell of *Arabidopsis* is polar. The nucleus and most of the cytoplasm lies at the end closest to the centre of the embryo sac, while the region of the egg near the micropyle is highly vacuolated (see Fig. 1.1 for the structure of the ovule and embryo sac).

After fertilization, the *Arabidopsis* zygote elongates about threefold and then divides transversely to create the two-cell 'proembryo' (Fig. 4.4). The division converts the polarity of the zygote into the longitudinal axis of the proembryo. This first division is asymmetric, producing a large basal cell near to the micropyle and a small terminal cell close to the centre of the embryo sac (Fig. 4.4). Derivatives of the basal cell form the suspensor and the basal section of the root apical meristem, while derivatives of the terminal cell generate the remainder of the embryo proper.

Genetic studies suggest that the asymmetry of the zygotic division is not required to establish the longitudinal axis of the embryo proper. Mutations in the *GNOM* gene cause the zygote to expand without elongating and also to divide symmetrically, producing daughter cells of similar size (Fig. 4.4). The basal cell develops into a relatively normal, shortened suspensor but the development of the terminal cell into the embryo proper is highly abnormal. Cell divisions occur at abnormal orientations and most *gnom* embryos develop into a cone shape, with improperly formed, fused cotyledons that taper into a hypocotyl with no root. Despite these abnormalities, the longitudinal axis of cone-shaped *gnom* embryos is in the normal orientation.

Some *gnom* embryos are completely spherical, with no obvious signs of a longitudinal axis. However, even spherical *gnom* embryos usually possess a longitudinal axis, detectable as a polar pattern of gene expression (Fig. 4.4). In wild-type globular embryos, cells in the protoderm express *AtLTP1*, a gene encoding a lipid transfer protein believed to be involved in the formation of the cuticle. Initially, *AtLTP1* is expressed in all protoderm cells but its expression becomes restricted to the apical region, i.e. to the developing hypocotyl and cotyledons. This reflects cuticle formation by the shoot but not the root. *AtLTP1* therefore provides a marker of embryo polarity. Cone-shaped *gnom* embryos have a similar pattern of *AtLTP1* expression to the wild type. In ball-shaped *gnom* embryos, however, *AtLTP1* expression may remain uniform, indicating that no axis develops. More often, expression becomes restricted to the apical region of the embryo, indicating an axis with normal polarity, or to the basal region of the embryo, indicating an axis with reversed polarity.

The *GNOM* gene encodes a protein with similarity to yeast proteins involved in secretion. A possible explanation for the *gnom* phenotype is

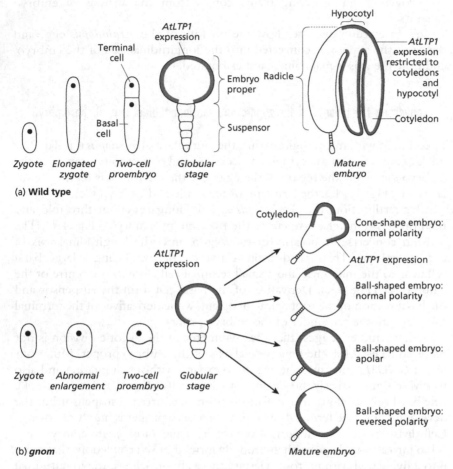

Fig. 4.4 Longitudinal axis development in wild-type and *gnom* embryos of *Arabidopsis*. In wild-type embryos (a), *AtLTP1* is expressed only in the apical epidermis (red). Consequently, the distribution of *AtLTP1* expression indicates the polarity of *gnom* embryos (b). Cone-shaped *gnom* embryos have normal polarity, whereas ball-shaped *gnom* embryos can have normal polarity, no polarity or reversed polarity. (Adapted from Mayer *et al.*, 1993; Vroemen *et al.*, 1996.)

that the GNOM protein is required to direct wall materials to the sites of cell wall deposition. If wall material were not directed accurately during cell division, then this could explain the abnormal division orientations in *gnom* embryos. The observation that embryo polarity can be expressed despite abnormal division planes suggests that there is a signal or gradient defining the apical–basal axis that is independent of the cellular architecture of the embryo. Presumably, the initial orientation of the signal or gradient depends on polarity inherited from the egg cell, which is normally perpetuated regardless of cell division patterns. In those ball-shaped *gnom* embryos that possess no longitudinal axis, there has been a failure in the maintenance of embryo polarity. The reversed longitudinal axes observed in some ball-shaped embryos suggest that after original embryo polarity has been lost, a new longitudinal axis can arise *de novo*.

Role of polar auxin transport in the embryo

Polar auxin transport is a prominent feature of the shoot-to-root axis (Fig. 4.5). In post-embryonic plants, auxin is produced at the shoot apex, predominantly in very young leaves. Auxin is then transported basipetally through the action of auxin efflux carriers located in the basal plasma membrane of vascular parenchyma cells. In the root there are two auxin streams. The shoot-to-root flow continues in the stele, but in addition, auxin is transported radially at the root tip and then back up the root in peripheral tissues—a pattern resembling an inverted fountain (Fig. 4.5). As discussed in later chapters, polar auxin flux provides both positional and physiological information, and has a variety of roles in the regulation of post-embryonic development.

Auxin transport along the longitudinal axis also appears to be a universal feature of higher plant embryos. From the globular stage onwards, auxin transport can be detected in the shoot-to-root direction. In the *Arabidopsis* embryo, this correlates with progressive polarization in the distribution of the PIN-FORMED1 (PIN1) protein, which is a component of an auxin efflux carrier (Fig. 4.6). In early embryos, PIN1 has an apolar distribution relative to the longitudinal axis. From the mid-globular stage onwards, however, PIN1 becomes increasingly restricted to the basal membranes

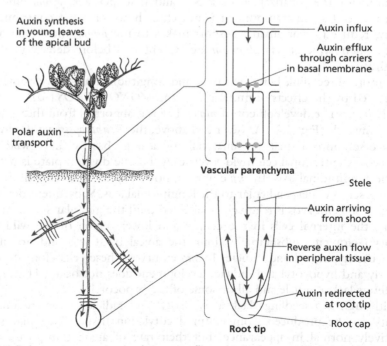

Fig. 4.5 Polar auxin transport in the pea seedling. Auxin is transported through vascular parenchyma cells from the shoot tip to the root tip. At the root tip, auxin is redistributed and transported back up the root in peripheral tissues. This pattern of transport appears to be universal in flowering plants.

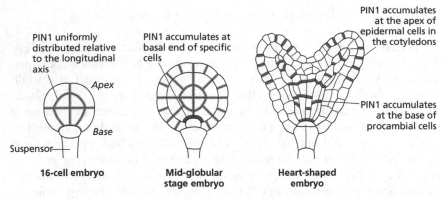

Fig. 4.6 The polarization of PIN1 distribution in the *Arabidopsis* embryo. The position of the PIN1 protein is indicated in red. (Adapted from Steinmann *et al.*, 1999.)

of presumptive procambial cells, consistent with the overall basipetal auxin flux. Interestingly, PIN1 also becomes localized to the apical membrane of epidermal cells in the cotyledons; the significance of this is discussed later.

The role of polar auxin transport in the development of the embryonic axes is not well understood. However, there is mounting evidence supporting its importance. Firstly, the distribution of PIN1 relative to the longitudinal axis remains random in *gnom* embryos, suggesting that *GNOM* is required for the polarization of PIN1, and that, perhaps, *gnom* embryos have reduced auxin transport. It is not clear, however, what contribution, if any, reduced polar auxin transport makes to the *gnom* phenotype. Certainly, phenotypic effects of *gnom* are evident well before defects in PIN1 localization.

A more direct link between auxin and longitudinal axis development is suggested by the effects of mutations in the *MONOPTEROS* (*MP*) gene of *Arabidopsis*. The development of *mp* embryos is abnormal from the two-cell stage onwards (Fig. 4.7). As described above, the *Arabidopsis* zygote divides transversely into a small terminal cell and a large basal cell. In wild-type embryos, the terminal cell divides vertically, i.e. the division plate is parallel to the longitudinal axis. In *mp* embryos, however, the terminal cell divides transversely, i.e. perpendicular to the longitudinal axis. Subsequent development of *mp* embryos is relatively unaffected until the globular stage. At this point, the internal cells in the central and lower region of the wild-type embryo elongate substantially along the apical–basal axis and go on to form the hypocotyl and radicle. In *mp* embryos, these cells elongate only slightly and hypocotyl and radicle development does not occur. Hence, *mp* seedlings lack a radicle and all or some of the hypocotyl.

Although *mp* seedlings lack a hypocotyl and radicle, complete mutant plants can be produced from cultured cotyledon tissue. Such plants are relatively normal in appearance, but their rate of auxin transport is significantly less than wild type. It is possible, therefore, that a reduction in polar auxin transport causes the partial failure of the longitudinal axis in *mp* embryos. The polarization of PIN1 distribution occurs normally in *mp*

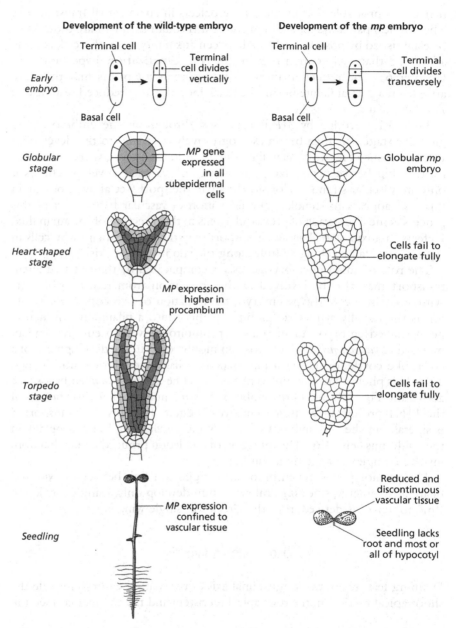

Fig. 4.7 *MP* expression and the effects of the *mp* mutation in *Arabidopsis* embryos. The regions of *MP* expression are coloured red (shown for the globular stage onwards) in the development of the wild-type embryo. (Adapted from Przemeck *et al.*, 1996; Hardtke & Berleth, 1998.)

embryos. Therefore, although auxin transport in *mp* mutants is probably reduced, it is likely that it occurs in the correct direction.

A relationship between *MP* and auxin has been confirmed by the cloning of the *MP* gene. *MP* encodes an Auxin Response Factor (ARF). ARFs bind to the promoters of auxin-inducible genes and regulate their transcription, hence it is probable that *mp* cells have defects in auxin signal transduction. The physical pathways for auxin movement through the plant are thought to be established by positive feedback between auxin flux through a cell, and the ability of that cell to transport auxin (the **canalization** hypothesis; see Chapter 6). The reduction in auxin transport in *mp* plants may therefore arise from a partial failure in this feedback loop due to a reduced response of *mp* cells to auxin.

The ARF encoded by *MP* is expressed throughout the embryo at the globular stage but then becomes progressively restricted to the developing vascular tissue (Fig. 4.7). As will be discussed in Chapter 6, there is a strong relationship between polarized auxin flux and vascular development. Not only are vascular strands major sites of auxin transport (see above), but auxin flux also appears to stimulate the formation of vascular tissue. Part of this process is the elongation of prevascular cells in the direction of the auxin flux. Defects in auxin signalling and transport may therefore explain why cells in *mp* embryos fail to elongate fully along the shoot-to-root axis.

The role of auxin in embryogenesis is complicated by the fact that auxin transport may also be involved in the conversion from radial to bilateral symmetry. In the wild-type embryo, the initiation of two cotyledons transforms the radially symmetric globular embryo into a bilaterally symmetric, heart-shaped embryo. Auxin transport inhibitors given to cultured Indian mustard (*Brassica juncea*) embryos commonly result in the development of a collar-like cotyledon giving a pin-shaped, radially symmetric embryo (Fig. 4.8). This phenotype is reminiscent both of cone-shaped *gnom* embryos (see above and Fig. 4.4) and of the embryos of *pin1* mutants. The distribution of the PIN1 protein in wild-type embryos suggests that auxin is transported basipetally in the procambium of developing cotyledons, but acropetally in the epidermis (Fig. 4.6). The initiation of cotyledon primordia may therefore involve complex changes in auxin flux.

The phenotype of *mp* embryos also suggests a link between auxin and bilateral symmetry, since *mp* embryos often develop only a single cotyledon ('monopteros' is derived from the Greek for 'single wing').

Apical meristem formation

Defining features of the longitudinal axis of the mature embryo include the shoot apical meristem, the root apical meristem and the vascular connection

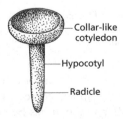

Collar-like cotyledon

Hypocotyl

Radicle

Fig.4.8 An Indian mustard embryo treated with a polar auxin transport inhibitor in tissue culture. The collar-like cotyledon makes the embryo radially symmetric. (Adapted from Hadfi *et al.*, 1998.)

between them. This section considers the mechanisms that position the apical meristems on the longitudinal axis.

Positioning the root apical meristem

The positioning of the embryonic root apical meristem (RAM) may also depend on auxin. Tissue culture experiments demonstrate that high ratios of auxin to the antagonistic hormone cytokinin promote root development, while low auxin to cytokinin ratios promote shoot development. It is possible that the apical–basal flow of auxin in the embryo results in a gradient of auxin concentration, with high concentrations in the presumptive root and low concentrations in the presumptive shoot. There is suggestive evidence that this concentration gradient either induces or modulates the development of the RAM.

Experiments on *Arabidopsis* seedlings transformed with an auxin-induced reporter gene suggest that the RAM is a site of high auxin concentration or response (Fig. 4.9). In the seedling root, expression of the reporter gene is strongest in the quiescent centre and columella (central root cap) of the RAM, with maximum expression in the columella initials. Furthermore, if the position of the RAM is experimentally changed, the position of the putative auxin maximum also changes. As discussed in case study 3.1, when the quiescent centre of the meristem is ablated, a new quiescent centre and columella are regenerated from former procambium cells. Performing this experiment on plants carrying the auxin-induced reporter gene reveals that as a functional RAM is re-formed, a new auxin maximum develops at the regeneration site.

There is some evidence that the auxin maximum is required for RAM development. For example, the absence of the RAM in *mp* embryos could result from reduced apical–basal auxin flux (see above), preventing the accumulation of auxin at the base of the embryo and therefore the formation of the RAM. To determine conclusively the role of auxin in RAM development, it will be necessary to study the genes required for meristem formation. Several mutants of *Arabidopsis* specifically lack a functional RAM and also fail to produce RAMs in tissue culture. A well-characterized example is *hobbit* (*hbt*).

Mutations in *HBT* affect the development of the RAM from the globular stage of embryogenesis onwards. In wild-type embryos, the quiescent centre and the central section of the root cap are derived from the **hypophysis**—the cell at the top of the suspensor. In *hbt* mutants, divisions in the hypophysis and its derivatives are abnormal giving a disordered pattern of cells in the quiescent centre and columella region of the meristem. Strong *hbt* mutants show no RAM activity. Discovering how, or whether, auxin regulates the expression of genes like *HBT* should reveal the role of auxin in positioning the embryonic RAM.

Experiments on plants carrying the auxin-induced reporter gene also relate the proposed auxin gradient to the polarity of the root. When plants carrying the reporter gene are treated with auxin transport inhibitors, reporter gene expression spreads to a broader area of the root tip. This suggests that auxin

Region of highest auxin-induced reporter gene expression

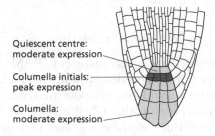

Quiescent centre:
moderate expression

Columella initials:
peak expression

Columella:
moderate expression

(a)

A bipolar *Arabidopsis* root resulting from prolonged exposure
of roots to polar auxin transport inhibitors

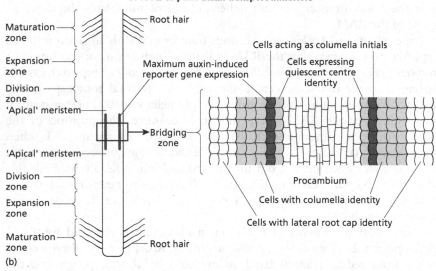

Maturation
zone

Expansion
zone

Division
zone

'Apical' meristem

Maximum auxin-induced
reporter gene expression

Cells acting as columella initials

Cells expressing
quiescent centre
identity

Root hair

'Apical' meristem

Bridging
zone

Division
zone

Expansion
zone

Maturation
zone

Root hair

Procambium

Cells with columella identity

Cells with lateral root cap identity

(b)

Fig. 4.9 The auxin maximum and the effects of polar auxin transport inhibitors on the *Arabidopsis* root. Cells shown in red display high expression of an auxin-induced reporter gene. (a) Reporter gene expression suggests a peak in auxin levels and/or response in the root apical meristem. (b) A bipolar *Arabidopsis* root generated by prolonged exposure of the root to polar auxin transport inhibitors. The ability of polar auxin transport inhibitors to bring about such pattern duplications suggests that auxin has a role in regulating pattern in the root tip. (Adapted from Sabatini *et al.*, 1999.)

is accumulating in a broader region of the root tip. It is likely that the transport inhibitors prevent all shoot-derived auxin from reaching the root, but that an established RAM acts as a site of auxin synthesis. In inhibitor-treated plants, such root-synthesized auxin would be trapped at the root tip, leading to the broader pattern of reporter gene expression.

Prolonged exposure to auxin transport inhibitors causes the transformation of the RAM into a bipolar structure (Fig. 4.9). The bipolar meristem continues to initiate root tissues in the original position, but it also initiates a new root in the direction of the former root cap. This second root has reversed polarity, as shown by the regions of cell division, cell elongation and cell maturation, and by the orientation of root hairs. The two roots are bridged by a region that contains vascular tissue, cells with quiescent centre identity, and cells with root cap identity (Fig. 4.9). Reporter gene expression indicates that the bridging region is the site of highest auxin concentration or response, suggesting that the bipolar root possesses a bipolar auxin gradient.

Positioning the shoot apical meristem

As will be discussed in Chapter 9, several genes necessary for shoot apical meristem (SAM) function have been identified by mutations that disrupt the activity and/or structure of the SAM. Prominent among these are *SHOOT-MERISTEMLESS* (*STM*), *WUSCHEL* (*WUS*), *CLAVATA1* (*CLV1*) and *CLAVATA3* (*CLV3*). These genes display characteristic patterns of expression in the SAM that develop progressively during embryogenesis. The first indication of SAM development is the expression of *WUS* in cells at the apex of the 16–cell embryo, i.e. the very early globular stage. Subsequently, *STM* is expressed in the apex of the late globular embryo, and *CLV1* and *CLV3* are expressed at the site of the presumptive SAM in the early heart-shaped embryo. The SAM becomes histologically distinguishable at the torpedo stage, and as embryogenesis proceeds to this point, the expression of each of the four genes becomes progressively restricted to a characteristic region of the meristem. Gene expression patterns in the SAM are described in detail in case study 9.1.

Mutational analysis demonstrates that after embryogenesis, the expression of *STM*, *WUS*, *CLV1* and *CLV3* are mutually interdependent (see case study 9.1). For example, complete loss-of-function *stm* mutants germinate without a SAM and mutant seedlings lack *WUS*, *CLV1* and *CLV3* expression at the shoot apex. In contrast, the initial expression of each of the four genes in the embryo appears to be somewhat independent of the other three genes. Loss of function of any one gene in the group does not prevent the embryonic expression of the other members. This suggests that the mechanism of SAM initiation during embryogenesis differs significantly from the mechanism of SAM maintenance during post-embryonic development.

Presumably, the genes required for a functional SAM are activated in response to positional information in the embryo. Although the nature of this information is unknown, the effects of the *cup-shaped cotyledon1* (*cuc1*) and *cup-shaped cotyledon2* (*cuc2*) mutations suggest a relationship between SAM formation and cotyledon development. Single *cuc1* or *cuc2* mutants have an almost wild-type phenotype. However, *cuc1 cuc2* double mutants germinate with fused cotyledons, forming a cup shape, and have no SAM. Plants regenerated from such seedlings develop almost normally, suggesting that *CUC1* and *CUC2* are redundantly required for cotyledon separation and SAM formation in the embryo, but not for SAM maintenance during vegetative growth. *CUC1* and *CUC2* encode closely related proteins belonging to the NAC-domain family of putative transcription factors and are homologous to the *NO APICAL MERISTEM* (*NAM*) gene of *Petunia*, which is also required for both cotyledon separation and SAM formation.

Consistent with their redundant function, the *CUC* genes have similar expression patterns during embryogenesis. *CUC2* is first expressed in a stripe of cells across the apex of the globular embryo, between the sites of the

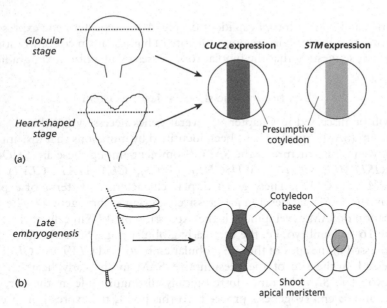

Fig. 4.10 The expression of *CUC2* and *STM* during embryogenesis. (a) During the globular and heart-shaped stages, both *CUC2* (red) and *STM* (pink) are expressed in a stripe at the apex of the embryo between the presumptive cotyledons. (b) Late in embryogenesis, *CUC2* and *STM* expression resolve into complementary patterns. *CUC2* is expressed between the cotyledon bases, and between the shoot apical meristem and the cotyledons. *STM* is expressed in the shoot apical meristem. (Adapted from Aida *et al.*, 1999.)

presumptive cotyledons. Interestingly, this is very similar to the initial pattern of *STM* expression, which occurs slightly later in the globular stage (Fig. 4.10). As embryogenesis proceeds, *CUC2* and *STM* expression resolve into complementary patterns. *STM* expression becomes restricted to the centre of the apex at the site of the presumptive SAM, whereas *CUC2* expression becomes restricted to the regions between the cotyledon primordia, and between the edge of the SAM and the bases of the cotyledons.

STM is not expressed in *cuc1 cuc2* double mutant embryos, indicating that *CUC1* and *CUC2* act to promote *STM* transcription at the embryonic apex. In *stm* mutant embryos, *CUC* gene expression is initially normal but becomes variable late in embryogenesis, suggesting feedback between SAM formation and *CUC* gene activity. Consistent with this, in addition to lacking a SAM, strong *stm* mutants have partially fused cotyledons.

Case study 4.3: Radial axis of the *Arabidopsis* embryo

Unlike the longitudinal axis, which pre-exists in the polarized *Arabidopsis* egg cell, the radial axis of the embryo becomes apparent only after several rounds of cell division. The first three rounds of division from the terminal cell of the two-cell embryo create a ball of eight cells above the developing suspensor. The eight cells are arranged in two tiers of four. The radial axis of the embryo becomes visible with the next round of cell division. Each of the eight cells divides periclinally (parallel with the surface of the embryo) and asymmetrically to produce a small outer protoderm cell and a large inner cell. In subsequent cell divisions, the group of inner cells differentiates into a central core of vascular tissue and an enveloping region of ground tissue (see Fig. 1.2).

There is good evidence that the radial axis in all parts of the plant is under similar control during both embryonic and post-embryonic development. For example, mutations in some genes have similar effects on the development of the radial axis in the radicle, hypocotyl, root and shoot. This case study will focus on the radicle and root as the phenotypes of radial pattern mutants are particularly well characterized in these organs. The case study considers radial axis formation. It then discusses the characteristics of positional information along the radial axis, and the relationship between the longitudinal and radial axes of the embryo.

Radial axis formation

During the development of the radicle, periclinal divisions in the vascular tissue create an outer pericycle and an inner core of xylem and phloem. Subsequently, the single layer of ground meristem cells subdivides into a layer of cortex and a layer of endodermis (Fig. 4.11). The radial pattern of the radicle matches the radial pattern of post-embryonic roots. There is an outer layer of epidermis; then single layers of cortex, endodermis and pericycle; and finally a central core of conductive tissue (cf. Fig. 1.5).

Mutations in several genes disrupt the radial pattern of the radicle. Loss of function of *WOODEN LEG* (*WOL*) and *GOLLUM* (*GLM*) both produce an abnormal pattern of vascular tissue. Interestingly, *WOL* encodes a cytokinin receptor, indicating a role for cytokinin in radial patterning. In contrast, mutations in *SHORT-ROOT* (*SHR*) and *SCARECROW* (*SCR*) disrupt the radial pattern of the ground tissue. The effects of the *wol, glm, shr* and *scr* mutations on the structure of the radicle are repeated in the primary root, in lateral roots and in roots regenerated from callus. These phenotypes suggest continuity in radial pattern formation and maintenance throughout the plant life cycle. This is a particularly striking conclusion since the radial pattern of the radicle develops by characteristic cell divisions occurring simultaneously along the basal region of the embryo, whereas the radial pattern of post-embryonic roots reflects divisions in a single group of initials in the root apical meristem (RAM).

The role of *SCR* has been studied in detail. In *scr* mutants, one layer of ground tissue is missing from the root due to the absence of key asymmetric divisions (Fig. 4.11). In the radicle, the *scr* phenotype arises because ground meristem cells along the radicle (and hypocotyl) of the torpedo-stage embryo fail to divide asymmetrically to produce an outer layer of large cortical cells and an inner layer of smaller endodermal cells. The defect in post-embryonic *scr* roots reflects the loss of specific asymmetric divisions in the apical meristem. The cortex and endodermis of post-embryonic roots both descend from a single ring of cortical initials (Fig. 4.11, see also Fig. 3.4). Each cortical initial divides in the transverse plane to produce a new cortical initial and a daughter cell. This daughter cell divides asymmetrically

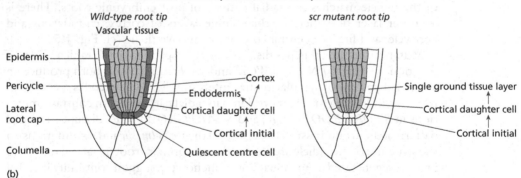

Fig. 4.11 The development of radial pattern in wild-type and *scr* mutant roots of *Arabidopsis*: cells expressing *SHR* are pink, cells expressing *SCR* are red. (a) Transverse sections through the developing radicle of wild-type and *scr* embryos. *SHR* expression is activated in the procambium of the globular embryo, and induces *SCR* expression in the ground tissue of the heart-shaped embryo. *SCR* is necessary for the periclinal division of the ground tissue into cortex and endodermis. In *scr* embryos, the ground tissue remains as a single layer. (b) Medial longitudinal sections through wild-type and *scr* root tips. In the wild-type root tip, *SHR* expression in the pericycle and vascular tissue induces *SCR* expression in the surrounding cell layer. *SCR* is necessary for the periclinal division of the cortical daughter cell, which generates the cortex and endodermis. In the *scr* root tip, the cortical daughter cell does not undergo a periclinal division and there is only one layer of ground tissue. (Adapted from Scheres *et al.*, 1995; Di Laurenzio *et al.*, 1996; Helariutta *et al.*, 2000.)

in the periclinal plane (parallel to the root surface) to produce an outer cortical cell and an inner endodermal cell. In *scr* mutants, however, the daughter cell divides transversely rather than periclinally, leaving the root with just one layer of ground tissue.

The single layer of ground tissue in *scr* roots displays characteristics of both the cortex and endodermis. The layer has a Casparian strip, which is an endodermal characteristic, but its cells are also recognized by an antibody that binds to cortical (and epidermal) but not endodermal cells. Therefore, although the *scr* mutation reduces the number of cell layers in the root, it does not appear to disrupt positional information along the radial axis. Instead, positional information defining different cell types along the radial axis is combined to give cells of a heterogeneous type.

The phenotype of *scr fass* double mutants confirms that the *scr* mutation does not delete positional information. As discussed in case study 3.3, *fass* mutants have randomly oriented cell divisions and unusually thick roots (see Fig. 3.9). *scr fass* double mutants also have unusually thick roots, showing that the *fass* mutation overcomes the reduction in cell layers caused by the *scr* mutation. The roots of *scr fass* double mutants have both a cortex and endodermis and so resemble the roots of *fass* single mutants.

The *SCR* gene has been isolated and its sequence suggests that it encodes a transcription factor. *SCR* is expressed in the ground tissue of embryos before the asymmetric division that creates the cortex and endodermis (Fig. 4.11). After the division, expression is restricted to the endodermis. In post-embryonic roots, *SCR* is expressed in the cortical initial–daughter cell pairs and the quiescent centre of the apical meristem (which is often disrupted in *scr* roots). In older tissues, *SCR* expression is restricted to the endodermis. The expression of *SCR* in both the embryonic and post-embryonic endo-dermis is particularly interesting because there is mounting evidence (dis-cussed below) that pattern formation in the post-embryonic root is regulated by information from older tissues higher up the root, rather than from within the RAM itself. If the pattern established in the radicle directs differ-entiation in the post-embryonic root, this could explain why mutations in a single gene can affect both processes.

Characteristics of positional information along the radial axis

A combination of genetic and laser ablation experiments suggests that devel-opment along the radial axis of the root requires both radial and longitudinal signalling. These are discussed in turn.

Radial signalling

The interaction between the *SHR* and *SCR* genes indicates that the patterning of the ground tissue in the root requires a signal from the stele. *SHR* encodes a putative transcription factor similar in sequence to that encoded by *SCR*. Furthermore, like *scr* mutants, *shr* mutants lack a layer of ground tissue. How-ever, whereas *scr* roots possess a single ground tissue layer with characteristics of both the cortex and endodermis (see above), *shr* roots simply lack an endoder-mis. *SCR* and *SHR* also have different expression patterns: *SCR* is expressed in the ground tissue (see above), whereas *SHR* is expressed in the stele (Fig. 4.11). Significantly, *scr* mutants have normal *SHR* expression (Fig. 4.11), but *shr* mutants have dramatically reduced *SCR* expression (not shown in Fig. 4.11). Therefore, *SHR* is required for the expression of *SCR* but not vice versa.

These data suggest that *SHR* expression in the stele mediates the production or transport of a signal to the ground tissue which allows *SCR* expression, division into two ground tissue layers, and endodermal development. It is probable that this signal is the SHR protein itself. Although the *SHR* gene is only expressed in the stele, the protein can be detected in the immediately adjacent layer of ground tissue. Presumably, the protein is transported to the ground tissue via plasmodesmata.

Interestingly, the shoots of *scr* and *shr* mutants also lack a layer of ground tissue—the shoot endodermis. This demonstrates similarities in the mechanisms controlling the radial pattern of roots and shoots.

Longitudinal signalling

Some important characteristics of positional information along the radial axis have been revealed by laser ablation experiments on the *Arabidopsis* RAM (Fig. 4.12). Firstly, such experiments suggest that information about radial position is transmitted from older tissues towards the root tip. If the daughter cell from a cortical initial–daughter cell pair is ablated, the development of the root is largely unaffected. The cortical initial below the ablated daughter produces a new daughter cell that divides asymmetrically in the usual way to produce endodermis and cortex. If, however, three adjacent daughter cells from cortical initial–daughter cell pairs are ablated, the cortical initial below the middle daughter produces daughters that do not divide asymmetrically into endodermis and cortex.

The result is a partial mimic of the *scr* mutation. The segment of the root above the middle cortical initial develops only a single column of ground tissue, suggesting that the ablation deprives that section of apical meristem of the positional information required for the asymmetric division into endodermis and cortex (Fig. 4.12). The ablated cells separate the meristem from older tissue, hence at least some of the information determining the radial pattern generated by the meristem appears to originate in older sections of the root. Since the transmission of information about radial position requires living cells, such information probably travels through the symplast.

The effects of ablating three adjacent cells lying across the root are different. If a daughter cell from a cortical initial–daughter cell pair is ablated together with the adjacent pericycle cell on its inside, and the epidermal cell on its outside, then the development of the root is not affected (Fig. 4.12). The cortical initial beneath the ablated daughter cell produces a new daughter that divides asymmetrically in the normal way. This confirms that the disruption to root development caused by ablating three adjacent daughter cells (above) does not represent a generalized response to root damage, instead it is a specific consequence of ablations within a single tissue layer.

The genetic data showing that radial axis development in the radicle and in post-embryonic roots occurs by the same mechanism are consistent with the flow of information about radial position from older tissues towards the root tip. Given that the primary root develops from the radicle, the radial pattern established in the embryo could direct the activity of the primary RAM. Similarly, cell divisions in an emerging lateral root occur along the length of

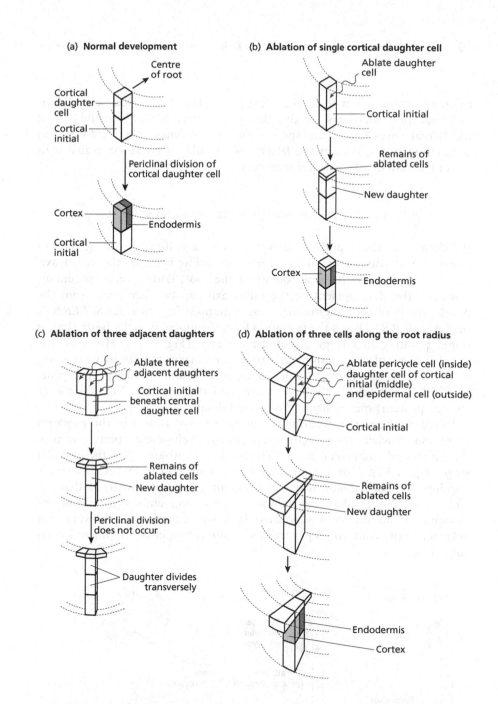

Fig. 4.12 Laser ablation experiments demonstrate that patterning at the root tip is mediated by signals from the mature root in *Arabidopsis*. The figure shows schematic, three-dimensional representations of the cortical initial and cortical daughter cell following various laser ablations. (a) During normal development, the cortical initial divides transversely (not shown) to produce a cortical daughter cell. The cortical daughter cell then divides periclinally to produce a cortical cell and an endodermal cell. (b) If a single cortical daughter cell is ablated, the next cortical daughter cell to arise follows the normal pattern of development. (c) If a row of three cortical daughter cells is ablated, however, the cortical daughter cell that subsequently arises beneath the central ablated cell fails to divide periclinally. This suggests that the periclinal division of the cortical daughter cell depends on signals from cells farther up the root. (d) If a cortical daughter cell and the two cells next to it along the root radius are ablated, the next cortical daughter cell to arise develops normally. Together with the results shown in (c), this suggests that the signal that induces the cortical daughter cell to divide periclinally travels through the ground tissue. (Adapted from van den Berg *et al.*, 1995.)

the root before a clear apical meristem is established. Detailed observations of lateral root initiation show that the radial pattern of tissues at the base of the lateral arises before the apical meristem becomes active. Information originating in the base of the lateral root would therefore be available to direct the activity of the apical meristem.

Relationship between the longitudinal and radial axes in the embryo

If information about position along the radial axis flows down the root to direct the activity of the RAM, then the maintenance of the radial axis depends on the longitudinal polarity of the root. Early in embryogenesis, however, the development of the radial axis can be uncoupled from the development of the longitudinal axis. Mutations in either *RASPBERRY1* or *RASPBERRY2* block the transition from the globular to the heart-shaped stage, resulting in enlarged, globular embryos (Fig. 4.13). However, the radial pattern of gene activity in *raspberry* embryos appears to be normal. Both wild-type and *raspberry* embryos express the lipid transfer protein gene, *AtLTP1*, specifically in the protoderm, and the gene for 2S2 albumin (a storage protein) specifically in the ground tissue.

Interestingly, as the development of the embryo proper in the *raspberry* mutants is blocked, the suspensor changes its developmental pattern to produce a second embryo proper which also fails to progress past the globular stage (Fig. 4.13). This suggests that correct embryo-proper development is required for normal suspensor development. A popular theory is that the embryo proper produces an inhibitor, preventing embryogenesis in the suspensor. This theory is supported by classic experiments showing that ablation of the embryo proper leads to embryo-proper development by the suspensor.

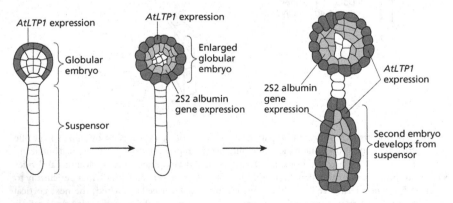

Fig. 4.13 The development of *raspberry Arabidopsis* embryos. The radial expression patterns of *AtLTP1* (red) and of the 2S2 albumin gene (pink) match those of wild-type embryos in both the enlarged, globular embryo of mutants and in the secondary embryo that develops from the mutant suspensor. (Adapted from Yadegari *et al.*, 1994.)

Conclusions

The radial and longitudinal axes of the embryo are apparent both in polar arrays of cell types along the axes and in the internal polarity of individual cells. Polarity in individual cells depends heavily on the cytoskeleton and is often converted to a multicellular axis through asymmetric division. Here a polarized cell divides perpendicular to the developing axis to produce daughters of different sizes that adopt different fates. For example, the *Arabidopsis* zygote divides asymmetrically to produce a small terminal cell from which most of the embryo proper develops, and a large basal cell from which the suspensor and a section of the root apical meristem develop. Consistent with the conclusions of Chapter 3, asymmetric divisions in plants do not result in lineage restrictions on cell fate. Suspensor cells, for example, switch to embryo-proper fates in *raspberry* mutants or after ablation of the embryo proper.

Early in angiosperm embryogenesis, the longitudinal and radial axes appear to develop independently. The longitudinal axis of *Arabidopsis* is apparent at the two-cell stage before the development of the radial axis has begun. Similarly, *raspberry* mutants are defective in longitudinal axis formation but produce a normal radial axis. Later in development, the two axes may not be independent: the radial axis appears to be patterned partly by the polar flow of information down the longitudinal axis.

Further reading

General

Brownlee, C. & Berger, F. (1995) Extracellular matrix and pattern in plant embryos: on the lookout for developmental information. *Trends in Genetics* 11, 344–348.

Goldberg, R.B., de Paiva, G. & Yadegari, R. (1994) Plant embryogenesis: zygote to seed. *Science* 266, 605–614.

Laux, T. & Jürgens, G. (1997) Embryogenesis: a new start in life. *Plant Cell* 9, 989–1000.

Case study 4.1: Longitudinal axis of the *Fucus* embryo

Establishing and maintaining the longitudinal axis

Alessa, L. & Kropf, D.L. (1999) F-actin marks the rhizoid pole in living *Pelvetia compressa* zygotes. *Development* 126, 201–209.

Kropf, D.L., Bisgrove, S.R. & Hable, W.E. (1999) Establishing a growth axis in fucoid algae. *Trends in Plant Science* 4, 490–494.

Kropf, D.L., Kloareg, B. & Quatrano, R.S. (1988) Cell wall is required for fixation of the embryonic axis in *Fucus* zygotes. *Science* 239, 187–190.

Quatrano, R.S. & Shaw, S.L. (1997) Role of the cell wall in the determination of cell polarity and the plane of cell division in *Fucus* embryos. *Trends in Plant Science* 2, 15–21.

Shaw, S.L. & Quatrano, R.S. (1996) The role of targeted secretion in the establishment of cell polarity and the orientation of the division plane in *Fucus* zygotes. *Development* 122, 2623–2630.

Role of the cell wall in elaborating the axis

Berger, F., Taylor, A. & Brownlee, C. (1994) Cell fate determination by the cell wall in early *Fucus* development. *Science* **263**, 1421–1423.

Cell wall components in flowering plants

Dénarié, J. & Cullimore, J. (1993) Lipo-oligosaccharide nodulation factors: a new class of signalling molecules mediating recognition and morphogenesis. *Cell* **74**, 951–954.

John, M., Röhrig, H., Schmidt, J., Walden, R. & Schell, J. (1997) Cell signalling by oligosaccharides. *Trends in Plant Science* **2**, 111–115.

Majewska-Sawka, A. & Nothnagel, E.A. (2000) The multiple roles of arabinogalactan proteins in plant development. *Plant Physiology* **122**, 3–9.

Case study 4.2: Longitudinal axis of the *Arabidopsis* embryo

GNOM

Mayer, U., Büttner, G. & Jürgens, G. (1993) Apical–basal pattern formation in the *Arabidopsis* embryo: studies on the role of the *gnom* gene. *Development* **117**, 149–162.

Steinmann, T., Geldner, N., Grebe, M. *et al.* (1999) Coordinated polar localization of auxin efflux carrier PIN1 by GNOM ARF GEF. *Science* **286**, 316–318.

Vroemen, C.W., Langeveld, S., Mayer, U. *et al.* (1996) Pattern formation in the *Arabidopsis* embryo revealed by position-specific lipid transfer gene expression. *Plant Cell* **8**, 783–791.

Polar auxin transport

Cooke, T.J. & Cohen, J.D. (1993) The role of auxin in plant embryogenesis. *Plant Cell* **5**, 1494–1495.

Hadfi, K., Speth, V. & Neuhaus, G. (1998) Auxin-induced developmental patterns in *Brassica juncea* embryos. *Development* **125**, 879–887.

Liu, C-M., Xu, Z-H. & Chua, N-H. (1993) Auxin polar transport is essential for the establishment of bilateral symmetry during early plant embryogenesis. *Plant Cell* **5**, 621–630.

Sabatini, S., Beis, D., Wolkenfelt, H. *et al.* (1999) An auxin-dependent distal organizer of pattern and polarity in the *Arabidopsis* root. *Cell* **99**, 463–472.

MONOPTEROS

Hardtke, C.S. & Berleth, T. (1998) The *Arabidopsis* gene *MONOPTEROS* encodes a transcription factor mediating embryo axis formation and vascular development. *EMBO Journal* **17**, 1405–1411.

Przemeck, G.K.H., Mattsson, J., Hardtke, C.S., Sung, Z.R. & Berleth, T. (1996) Studies on the role of the *Arabidopsis* gene *MONOPTEROS* in vascular development and plant axialization. *Planta* **200**, 229–237.

HOBBIT

Willemsen, V., Wolkenfelt, H., de Vrieze, G., Weisbeek, P. & Scheres, B. (1998) The *HOBBIT* gene is required for formation of the root meristem in the *Arabidopsis* embryo. *Development* **125**, 521–531.

Shoot apical meristem development

Aida, M., Ishida, T. & Tasaka, M. (1999) Shoot apical meristem and cotyledon formation during *Arabidopsis* embryogenesis: interaction among the *CUP-SHAPED COTYLEDON* and *SHOOT MERISTEMLESS* genes. *Development* **126**, 1563–1570.

Bowman, J.L. & Eshed, Y. (2000) Formation and maintenance of the shoot apical meristem. *Trends in Plant Science* **5**, 110–115.

Takada, S., Hibara, K., Ishida, T. & Tasaka, M. (2001) The *CUP-SHAPED COTYLEDON1* gene of *Arabidopsis* regulates shoot apical meristem formation. *Development* **128**, 1127–1135.

Further references concerning shoot apical meristem development are listed in Chapter 9.

Case study 4.3: Radial axis of the *Arabidopsis* embryo

SCARECROW (and other genes implicated in radial pattern)

Di Laurenzio, L., Wysocka-Diller, J., Malamy, J.E. *et al.* (1996) The *SCARECROW* gene regulates an asymmetric cell division that is essential for generating the radial organization of the *Arabidopsis* root. *Cell* **86**, 423–433.

Helariutta, Y., Fukaki, H., Wysocka-Diller, J. *et al.* (2000) The *SHORT-ROOT* gene controls radial patterning of the *Arabidopsis* root through radial signaling. *Cell* **101**, 555–567.

Nakajima, K., Sena, G., Nawy, T. & Benfey, P.N. (2001) Intercellular movement of the putative transcription factor SHR in root patterning. *Nature* **413**, 307–311.

Scheres, B., Di Laurenzio, L., Willemsen, V. *et al.* (1995) Mutations affecting the radial organisation of the *Arabidopsis* root display specific defects throughout the embryonic axis. *Development* **121**, 53–62.

Wysocka-Diller, J.W., Helariutta, Y., Fukaki, H., Malamy, J.E. & Benfey, P.N. (2000) Molecular analysis of SCARECROW function reveals a radial patterning mechanism common to root and shoot. *Development* **127**, 595–603.

Laser ablation

Van den Berg, C., Willemsen, V., Hage, W., Weisbeek, P. & Scheres, B. (1995) Cell fate in the *Arabidopsis* root meristem determined by directional signalling. *Nature* **378**, 62–65.

RASPBERRY1 and RASPBERRY2

Yadegari, R., de Paiva, G.R., Laux, T. *et al.* (1994) Cell differentiation and morphogenesis are uncoupled in *Arabidopsis raspberry* embryos. *Plant Cell* **6**, 1713–1729.

CHAPTER 5

Axis development in the leaf and flower

This chapter considers the axes of leaves and flowers. Leaves, cotyledons, bracts and the floral organs—petals, sepals, stamens and carpels—are **lateral organs**. It is probable that all lateral organs are derived from leaves and, therefore, that similar mechanisms generate axial information in each case. The flower as a whole represents a modified shoot, producing floral organs instead of leaves and usually following a determinate pattern of development.

Chapter 4 discussed the primary radial and longitudinal axes of the plant. The axes of the leaf are ultimately derived from these primary axes. During normal development, leaves display consistent orientation and polarity relative to the shoot. This suggests that axial information in the leaf does not arise *de novo* but depends on existing axial information. Similarly, because the flower is a modified shoot, the mechanisms that generate axial information in the floral meristem are likely to be similar to those controlling axial information in vegetative shoot apical meristems.

Leaves

The angiosperm leaf is almost always a determinate organ and usually projects out from the stem. Simple (as opposed to compound) leaves typically have three axes of asymmetry (Fig. 5.1): a **proximodistal axis**, leading from the base of the leaf to the tip; an **adaxial–abaxial axis**, leading from the upper to the lower epidermis; and a duplicated **centrolateral axis**, leading from the midrib to the margin. The acquisition and elaboration of the adaxial–abaxial axis, and its relationship to the centrolateral axis have been studied in some detail. These experiments are considered in case study 5.1. Less is known about the development of the proximodistal axis, which is discussed in case study 5.2. Last in this section, case study 5.3 considers the determinate nature of leaf development and the control of leaf size.

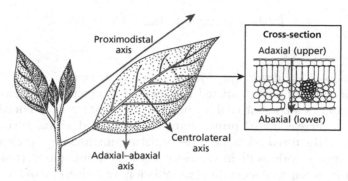

Fig. 5.1 The three axes of the leaf.

As discussed in Chapter 1, the terminology used to describe leaf development relates to the developmental rather than actual age of the leaf. The youngest visible leaf primordium is called **P1**, the next oldest is **P2**, and so on. Future leaf primordia are called 'incipient' and are numbered **I1, I2**, etc., with I1 being the first to appear. Hence I1 is the site at which the next leaf primordium will emerge from the shoot apical meristem. The length of time between the initiation of successive nodes is a **plastochron**.

Case study 5.1: Adaxial–abaxial axis of the leaf

In most species, mature leaves project out from the stem and so have an upper and lower surface. When the leaf primordium emerges from the shoot apical meristem, the face that will become the upper surface points towards the centre of the meristem and is called **adaxial**. The face that will become the lower surface points away from the meristem centre and is called **abaxial**. Consequently, the axis running from the upper to the lower surface of the mature leaf is the adaxial–abaxial axis.

Adaxial–abaxial (sometimes called dorsoventral) asymmetry in the leaf is apparent in the pattern of cell types along the adaxial–abaxial axis and in differential growth between the adaxial and the abaxial halves of the primordium. For example, the adaxial and abaxial epidermis usually possess differently shaped cells and have different patterns and/or types of specialized structures (such as stomata and trichomes). Adaxial–abaxial asymmetry also exists in leaf vascular bundles, xylem is adaxial and phloem is abaxial. Leaves project out from the stem because growth in the adaxial half of the primordium outstrips growth in the abaxial half.

This case study considers the acquisition of adaxial–abaxial asymmetry by the leaf primordium—a problem that has been tackled using both surgical and genetic experiments. The case study then discusses the maintenance of the adaxial–abaxial axis throughout leaf development. Finally, the relationships between the adaxial–abaxial axis and the centrolateral axis, and between adaxial leaf tissue and shoot apical meristem development are considered.

Acquisition of adaxial–abaxial asymmetry

Timing

Observation of leaf development in tobacco and *Arabidopsis* shows that the dicot leaf primordium is initiated as a radially symmetric outgrowth that rapidly acquires adaxial–abaxial asymmetry. The tobacco P1 primordium is cylindrical whereas the P2 primordium has a flattened adaxial surface (see Fig. 1.7). As discussed below, adaxial–abaxial asymmetries in gene expression patterns arise as early as P1 in the *Arabidopsis* leaf primordium, confirming that polarity is acquired soon after leaf initiation. In maize, one of the earliest indicators occurs at P3 when a distinct ligule-forming region can be seen on the adaxial leaf surface, indicating that polarity must arise before this stage (Fig. 1.8 illustrates leaf development in maize).

Basis of adaxial–abaxial asymmetry

Surgical experiments performed on several dicot species indicate that adaxial–abaxial polarity in the leaf depends on the radial axis of the shoot apical meristem (Fig. 5.2). An incision that isolates I1 from the centre of the meristem results in the outgrowth of a radially symmetric leaf that consists of an outer layer of abaxial-type epidermis surrounding a uniform ring of parenchyma and a central core of vascular tissue. The most likely interpretation is that the central region of the meristem produces a chemical signal that promotes adaxial leaf fates and orients the adaxial–abaxial axis. The incision prevents the primordium from receiving the signal, resulting in radial leaves displaying only abaxial fates.

The acquisition of adaxial–abaxial asymmetry has also been studied genetically. The *PHANTASTICA* gene of *Antirrhinum*; and *PINHEAD* (also called *ZWILLE*), *ARGONAUTE1*, *PHABULOSA* and the *YABBY* family genes of *Arabidopsis* are all implicated in the process.

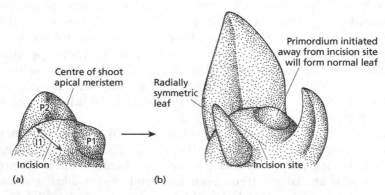

Fig. 5.2 The isolation of a leaf primordium from the meristem centre induces radial symmetry. If an incision is made between I1 and the centre of the shoot apical meristem (a), a radially symmetric leaf develops (b). (Adapted from Steeves & Sussex, 1989.)

PHANTASTICA

The *PHANTASTICA* (*PHAN*) gene of *Antirrhinum* is required for the acquisition of an adaxial–abaxial axis by the leaf. Loss-of-function *phan* mutants develop leaves with variable loss of adaxial–abaxial asymmetry, as revealed by the replacement of adaxial cell types with abaxial types. In the most severely affected leaves, adaxial–abaxial polarity is entirely lost and there is also no development of the lamina: the leaves are radially symmetric and needlelike (Fig. 5.3).

PHAN encodes a MYB-type transcription factor. It is expressed in the apical meristem at the future sites of leaf initiation (as early as I4 during the development of bracts on the inflorescence), and in leaf primordia up until the P3 stage. Throughout this period, *PHAN* expression is uniform along the adaxial–abaxial axis. This indicates that *PHAN* does not itself provide adaxial–abaxial information. One model is that the primordium requires *PHAN* expression to be able to respond to the polarizing signal produced by the apical meristem. This would explain why *phan* leaves are 'abaxialized', resembling the symmetric leaves produced when I1 is surgically isolated from the meristem centre (see above).

PINHEAD and *ARGONAUTE1*

The *PINHEAD* (*PNH*) and *ARGONAUTE1* (*AGO1*) genes of *Arabidopsis* are required for the development of adaxial leaf tissue. *PNH* was initially identified by analysis of mutants with abortive shoot apical meristem development (see below). *pnh* single mutants have normal leaves, and *ago1* single mutants have partially abaxialized lateral organs. However, plants with the genotype *ago1/ago1 pnh/+*—plants homozygous for *ago1* and heterozygous for *pnh*—have filamentous leaves with a severe reduction in adaxial characteristics. Therefore loss of some *PNH* function makes the phenotype of *ago1* much more severe, suggesting that *PNH* and *AGO1* act redundantly to allow or direct the development of adaxial tissue. The genes both encode proteins with similarity to eukaryotic translation initiation factors, however the biochemical functions of the PNH and AGO1 proteins are unknown.

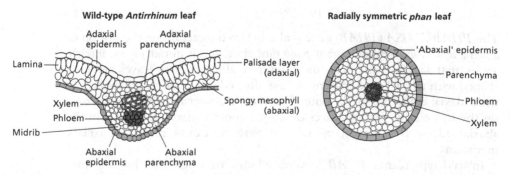

Fig. 5.3 Cross-sections through a wild-type *Antirrhinum* leaf and a radially symmetric *phan* leaf. (Adapted from Waites & Hudson, 1995.)

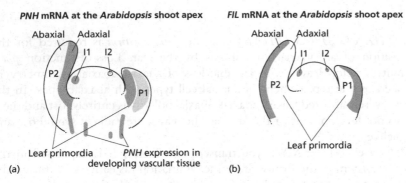

PNH mRNA at the *Arabidopsis* shoot apex

Abaxial Adaxial

I1 I2

P2

P1

Leaf primordia   *PNH* expression in
                     developing vascular tissue
(a)

FIL mRNA at the *Arabidopsis* shoot apex

Abaxial Adaxial

I1 I2

P2

P1

Leaf primordia

(b)

Fig. 5.4 *PNH* and *FIL* expression during leaf initiation. The distribution of *PNH* and *FIL* mRNA, respectively, is indicated in shades of red. (a) *PNH* is expressed uniformly across the I1 site and becomes restricted to adaxial leaf tissue by the P2 stage. *PNH* is also expressed in developing vascular tissue. (b) *FIL* is expressed uniformly across the I2 and I1 sites, and is restricted to abaxial leaf tissue by the P1 stage. Note that in a real *Arabidopsis* apex, the leaf primordia form a spiral so that I2, I1, P1 and P2 cannot be captured in a single cross-section. ((b) adapted from Siegfried *et al.*, 1999.)

Despite the apparent redundancy between *AGO1* and *PNH*, the genes have different expression patterns. *AGO1* is expressed ubiquitously in the plant, whereas the expression of *PNH* relates to the acquisition of the adaxial–abaxial axis in the leaf (Fig. 5.4a). *PNH* is expressed uniformly across the I1 site. At P1, *PNH* expression is higher in the adaxial half of the primordium, and by P2, *PNH* is only expressed in the adaxial region. The change in *PNH* expression may be a response to the proposed polarizing signal from the centre of the meristem. *PNH* is also expressed in vascular tissue in the stem (Fig. 5.4a). Its function there is unknown.

Interestingly, in addition to its developmental role, *AGO1* is required for the phenomenon of post-transcriptional gene silencing, in which the expression of some viral genes and transgenes is blocked by the degradation of their mRNAs. It is possible that the *ago1* leaf phenotype is caused by abnormalities in gene silencing; however, other *Arabidopsis* mutants with impaired gene silencing develop normally.

PHABULOSA

The *PHABULOSA* (*PHAB*) gene is also believed to act in the promotion of adaxial leaf fates. Semidominant *phab* mutations result in radially symmetric leaves that lack laminas. The most severely affected *phab* leaves are rod-shaped with adaxial-type epidermis. Less affected leaves are trumpet-shaped, with adaxial epidermis on the outer face of the trumpet and abaxial epidermis on the inner face. Mutant leaves develop ectopic axillary meristems on the abaxial side so that, in extreme cases, a *phab* leaf can be ringed by axillary meristems.

In wild-type plants, *PHAB* is expressed uniformly across I1 but becomes restricted to the adaxial region of the leaf by P2, a pattern similar to that of *PNH* (see above and Fig. 5.4a). Like *PNH*, *PHAB* is also expressed in vascular tissue. Unlike *PNH*, however, the expression of *PHAB* in I1 and P1 extends beyond the developing leaf in stripes leading to the centre of the meristem—giving a pattern reminiscent of a clock face. This intriguing observation may relate to the polarization of the leaf by a signal from the shoot apical meristem.

PHAB encodes a putative transcription factor containing DNA-binding motifs belonging to both the homeodomain and leucine zipper classes. The PHAB protein also contains a predicted sterol/lipid-binding domain, suggesting that its activity may be regulated by a sterol/lipid ligand. The *phab* mutations all affect this putative sterol/lipid-binding domain and are predicted to result in a constitutively active form of PHAB. Interestingly, unlike the pattern in wild-type plants, *PHAB* is expressed throughout the developing leaf in *phab* mutants. Together with the mutant phenotype, these data suggest that *PHAB* promotes adaxial leaf fates and that its expression is restricted to the adaxial region of the leaf by a feedback mechanism involving the polarized distribution of a sterol/lipid ligand. In wild-type plants, the putative ligand activates the PHAB transcription factor preferentially in adaxial regions. Activated PHAB then promotes adaxial cell fates and maintains *PHAB* expression in those regions. In *phab* mutants, however, constitutively active PHAB maintains *PHAB* expression and promotes adaxial cell fates throughout the leaf independently of the ligand.

The *YABBY* gene family

In direct contrast to *PNH* and *AGO1*, members of the *YABBY* gene family are required for the development of abaxial leaf tissue in *Arabidopsis*. The family contains six or seven related genes which encode putative transcription factors. In all the cases studied, changes in *YABBY* family expression affect aspects of adaxial–abaxial polarity in lateral organs. Three members of the family, *FILAMENTOUS FLOWER* (*FIL*), *YABBY2* (*YAB2*) and *YABBY3* (*YAB3*), have been studied in detail with respect to leaf development.

The three genes have similar expression patterns in the leaf (Fig. 5.4b). *FIL*, *YAB2* and *YAB3* expression begins at I2 in subepidermal cells. At this stage, the three genes are expressed uniformly relative to the future adaxial–abaxial axis of the leaf. At P1 expression becomes restricted to the abaxial side, including the abaxial epidermis and the future spongy mesophyll, a pattern roughly complementary to that of *PNH*. The abaxial pattern of expression continues through leaf development, until the transcripts of all three genes disappear in the mature leaf.

A functional relationship between *FIL* and *YAB3* expression and abaxial leaf fates has been demonstrated using transgenic plants and mutants. Transgenic embryos that constitutively express *FIL* or *YAB3* often fail to develop into viable seedlings. If they do germinate, the seedlings produced have narrow leaves with abaxial cell types in the adaxial epidermis. Therefore, ectopic expression of *FIL* or *YAB3* can induce ectopic abaxial characteristics. The available *fil* and *yab3* single mutants do not affect leaf development (*FIL* was identified because mutants develop filamentous flowers). However, *fil yab3* double mutants produce narrow leaves in which the abaxial epidermis has a mixture of adaxial and abaxial characteristics. Loss of *FIL* and *YAB3* function can therefore cause ectopic adaxial characteristics in abaxial sites.

In the model of leaf development in which a signal from the meristem centre promotes adaxial leaf fates, that signal would be predicted to inhibit, directly or indirectly, *FIL* and *YAB3* expression in adaxial tissue.

Maintenance of the adaxial–abaxial axis

As discussed in Chapter 3, cell-extrinsic, fate-determining information is present throughout leaf development. When a clone of cells with an epidermal lineage is placed into the palisade layer by a spontaneous periclinal division, the displaced cells always adopt fates appropriate to their palisade position (see case study 3.2). This happens regardless of how late in development the displacement occurs. Hence information defining position along the adaxial–abaxial axis is maintained until the final differentiation of leaf cells.

Although a signal from the centre of the apical meristem appears necessary to establish the adaxial–abaxial axis (see above), the mechanism that maintains the axis is probably intrinsic to the leaf. Dicot leaf primordia removed from the meristem at the P2 stage grow into fully developed, albeit small, leaves in tissue culture. By implication, all axes are autonomous to the leaf at this point.

Little is known about the maintenance of the adaxial–abaxial axis; however, analysis of the *lam1* mutation of tobacco indicates that signals conferring adaxial cell fates occur throughout the development of the blade.

LAM1

The phenotype of *lam1* tobacco mutants suggests that *LAM1* is needed specifically for the maintenance of the adaxial–abaxial axis (Fig. 5.5). Leaf primordia of *lam1* plants are initially indistinguishable from wild type. Like wild-type primordia, *lam1* primordia undergo adaxial–abaxial flattening at the P2 stage, suggesting that the acquisition of adaxial–abaxial polarity is unaffected by the mutation. However, although *lam1* leaves initiate a lamina, adaxial cell types in the lamina (i.e. adaxial epidermal cells and the palisade layer) are replaced by abaxial cell types. Furthermore, the lamina fails to grow along the centrolateral axis, a phenomenon discussed further below. The *lam1* mutation is recessive and so probably represents a loss of function. This would imply that *LAM1* is needed for the adaxial–abaxial axis of the leaf blade and to allow full blade growth. The DNA sequence and the expression pattern of *LAM1* are unknown.

The function of *LAM1* has been investigated further through the use of periclinal chimeras containing *lam1* and wild-type cells (Fig. 5.5). *lam1*–wild type–wild type plants (i.e. plants with a *lam1* L1, and a wild-type L2 and L3: see Chapter 3 for a description of periclinal chimeras) develop leaves

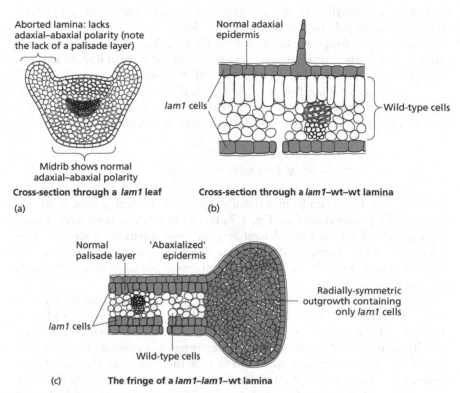

Aborted lamina: lacks adaxial–abaxial polarity (note the lack of a palisade layer)

Normal adaxial epidermis

lam1 cells

Wild-type cells

Midrib shows normal adaxial–abaxial polarity

Cross-section through a *lam1* leaf

(a)

Cross-section through a *lam1*–wt–wt lamina

(b)

Normal palisade layer

'Abaxialized' epidermis

Radially-symmetric outgrowth containing only *lam1* cells

lam1 cells

Wild-type cells

(c)   **The fringe of a *lam1*–*lam1*–wt lamina**

Fig. 5.5 Cross-sections through leaves of *lam1* tobacco plants and *lam1*/wild-type chimeras. For the chimeric plants, *lam1* cells are coloured red, and wild-type cells are white. (a) In *lam1* mutants, the development of the leaf midrib is normal but the leaf lamina aborts and lacks adaxial–abaxial polarity. (b) The leaves of *lam1*–wt–wt chimeras develop normally, producing a lamina with the correct sequence of cell layers. (b) The leaves of *lam1*–*lam1*–wt chimeras have abaxialized epidermis on their adaxial surface and produce radially symmetric outgrowths at the lamina fringe. These results suggest that wild-type cells produce a signal that confers adaxial identity on immediately adjacent, adaxially positioned *lam1* cells. (Adapted from McHale, 1992; McHale & Marcotrigiano, 1998.)

that are almost indistinguishable from wild type. Although the adaxial epidermis of such leaves is composed of *lam1* cells, these cells adopt adaxial, i.e. wild-type, fates. This suggests that *LAM1* expression in the L2- and L3-derived mesophyll generates a non-cell-autonomous signal that confers adaxial fates on the mutant adaxial epidermis.

lam1–*lam1*–wild type chimeras (*lam1* cells in the L1 and L2, and wild-type cells in the L3) have leaves that possess an approximately wild-type-sized blade but that have adaxial–abaxial abnormalities. Through most of the blade, *lam1* cells are confined to the epidermis and subepidermal layers (this is consistent with normal lineage patterns, see Chapter 3). These regions of the leaf blade develop a normal palisade layer showing that the core of wild-type mesophyll is sufficient to confer adaxial fates on *lam1* cells in the palisade position. However, the adaxial epidermis of the leaf blade has abaxial characteristics. The proposed 'adaxializing' signal produced by wild-type mesophyll cells is therefore short range and does not extend through the palisade layer to the upper epidermis.

Lastly, at the fringe of *lam1–lam1*–wild type leaves, regions of purely mutant tissue exist. Again, this is consistent with the normal lineage patterns of a dicot leaf in which the fringe of the blade consists of only L1 and L2 cells. The fringe of a *lam1–lam1*–wild type leaf produces radially symmetric, 'abaxialized' outgrowths, confirming that the inner core of wild-type L3 is required to maintain the abaxial–adaxial axis in such leaves.

Adaxial–abaxial asymmetry and lamina development

In dicots, the transition from a radially symmetric P1 leaf primordium to a flattened P2 primordium (see Fig. 1.7) results in bilateral symmetry. This is the stage at which the centrolateral axis becomes apparent. There is strong evidence that the extension of the lamina along the centrolateral axis requires the juxtaposition of adaxial and abaxial cell types. *phan* mutants (see above) provide a good example.

phan mutants are temperature-sensitive with more severe phenotypes at lower temperatures. At 25°C, *phan* leaves appear almost entirely wild type. At 17°C, almost all leaves completely lack an adaxial–abaxial axis and are radially symmetric. At intermediate temperatures, however, many *phan* leaves have laminas. Close examination shows that the adaxial epidermis of such laminas contains patches of abaxial cell types: demonstrated by the presence of stomata and characteristic cell shapes (Fig. 5.6). Sections through affected leaves indicate that the 'abaxial' patches in the adaxial epidermis occur in sections of the lamina that lack an adaxial–abaxial axis. Beneath each patch of ectopic 'abaxial' epidermis, the palisade layer fails to develop and the internal tissue consists only of spongy mesophyll.

Interestingly, each patch of ectopic abaxial epidermis is bounded by a ridge projecting up out of the plane of the leaf (Fig. 5.6). In cross-section,

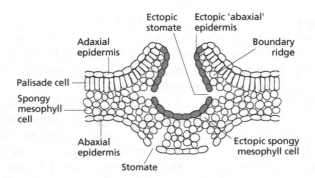

Fig. 5.6 Ectopic abaxial tissue on a *phan* leaf. Ectopic 'abaxial' epidermis is coloured red. The boundary ridges that surround the ectopic 'abaxial' epidermis appear to represent secondary leaf laminas, and are produced where adaxial and abaxial tissues are juxtaposed. (Adapted from Waites & Hudson, 1995.)

these boundary ridges resemble secondary leaf laminas. The ridges have palisade mesophyll beneath the face covered by adaxial epidermis, and spongy mesophyll beneath the face covered by the ectopic abaxial epidermis. Since the ridges occur at the junction between adaxial and abaxial cell types, they suggest that the juxtaposition of adaxial and abaxial cells induces lamina formation.

The recessive *leafbladeless1* (*lbl1*) mutation of maize results in a similar range of leaf phenotypes to those of *phan* mutants and strongly supports the juxtaposition hypothesis. The most severe *lbl1* leaf phenotype is a radially symmetric, thread-like leaf with abaxial-type epidermis. Less severely affected leaves develop reduced laminas. The adaxial epidermis on such laminas contains patches of abaxial tissue surrounded by ectopic, secondary laminas. The *lbl1* and *phan* phenotypes suggest that lamina initiation is induced by the juxtaposition of adaxial and abaxial cell types in both monocots and dicots. This may explain the frequent loss or reduction of the leaf lamina in mutants with abnormalities in adaxial–abaxial development.

Adaxial leaf tissue and shoot apical meristem development

There is a final aspect of the adaxial–abaxial axis to consider: its relationship to shoot apical meristem development. There is evidence that adaxial leaf tissue both promotes the formation of axillary meristems and maintains the development of the primary shoot apical meristem.

In a wild-type *Arabidopsis* leaf, an axillary meristem develops from adaxial cells at the leaf base. As described above, *phab* mutants develop either rod-or trumpet-shaped leaves that have an outer surface consisting of adaxial-type epidermis. Such leaves develop ectopic axillary meristems on the abaxial side so that, in extreme cases, a *phab* leaf can be ringed by axillary meristems. This suggests that the combination of adaxial and proximal cell fate is required to generate the meristem.

In addition to the relationship between adaxial cell types and axillary meristem initiation, there is also evidence to suggest that leaf adaxial tissue is required to maintain the primary shoot apical meristem. Firstly, as described earlier, the *phan* mutation of *Antirrhinum* is temperature-sensitive. At 15 °C, *phan* mutants arrest development at the shoot apical meristem. Given that *PHAN* is only expressed in leaf primordia, the effect of the mutation on the apical meristem is non-cell-autonomous, suggesting that the leaf primordium produces a signal that maintains meristem activity. Secondly, embryos that constitutively express either *FIL* or *YAB3* often fail to produce viable seedlings (see above). In such cases, the seedling shoot apical meristem has arrested. Hence the ectopic abaxial cell fates induced by constitutive *FIL* or *YAB3* expression can halt shoot apical meristem development. Lastly, as described above, *PNH* and *AGO1* function to specify adaxial fates in the leaf primordium. In *pnh* single mutants, the seedling shoot apical meristem arrests. In full *pnh ago1* double mutants (homozygous for both mutations) no shoot apical meristem develops in the embryo. Again, loss of adaxial leaf fates is associated with loss of shoot apical meristem function.

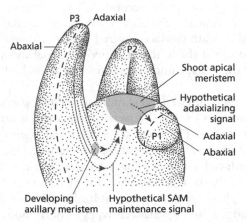

Fig. 5.7 The reciprocal relationship between the shoot apical meristem and adaxial leaf tissue. A hypothetical signal from the centre of the apical meristem induces adaxial–abaxial polarity in P1. Later (shown hypothetically for P3) a signal produced by adaxial leaf tissue maintains the shoot apical meristem.

All of these observations can be explained by a model in which adaxial tissue in the leaf primordium produces a signal that maintains shoot apical meristem activity. Given the evidence from surgical experiments that adaxial cell types are specified by a signal from the apical meristem, the shoot apical meristem and the leaf primordia it initiates may reciprocally regulate each other's development (Fig. 5.7).

Case study 5.2: Proximodistal axis of the leaf

In the dicots tobacco and *Arabidopsis*, vegetative leaves consist of a distal blade and a proximal petiole (Fig. 1.7 shows the development of the tobacco leaf). In maize, a monocot, the leaf consists of a distal blade and a proximal sheath (see Fig. 1.8). In all three species, proximodistal differences between leaf cells are visible at the P3 stage. Also in all these cases, the leaf matures in a tip-to-base (**basipetal**) wave (Chapter 1). Hence the proximodistal axis of the leaf is characterized by both spatial and temporal differences in cell fate.

Leaf primordia project out from the shoot apical meristem; therefore, relative to the longitudinal axis of the shoot, the tip of the leaf primordium is more 'apical' than the base. Similarly, auxin transport in the young leaf is assumed to be basipetal, as it is in the shoot. In these respects at least, the tip-to-base polarity of the leaf can be considered as an extension of the shoot-to-root polarity of the shoot. However, the mechanistic connections between the longitudinal axis of the shoot and the proximodistal axis of the leaf are unknown.

This case study uses the *knotted1* (*kn1*) mutant of maize to illustrate two models for the basis of proximodistal information in the leaf.

knotted1

The boundary between the sheath and the blade of the maize leaf is marked by a flap called the **ligule** (which keeps rain from running off the leaf blade into the nested sheaths below) and by two wedges of tissue called **auricles** (see Fig. 1.8 and Fig. 5.8). Dominant *kn1* mutations result in knots of tissue along the lateral veins of the leaf blade, i.e. those veins running up the blade parallel to the midrib (Fig. 5.8). The knots are formed by clusters of mesophyll cells that continue to divide after cell divisions in neighbouring regions of the blade have ceased. Mesophyll cells between the knots, and epidermal cells overlying the knots, develop sheath rather than blade characteristics. The result is that sheath tissue projects along the lateral veins up into the blade. Each region of ectopic sheath is accompanied near its apex by an ectopic ligule and ectopic auricles. The prolonged division of blade mesophyll cells to produce the 'knots' can also be interpreted as a sheath characteristic because cell divisions continue for longer in the sheath than in the blade. Hence the proximodistal axis of *kn1* leaves is disrupted.

KN1 encodes a homeodomain transcription factor and is the founding member of a family of genes known as *KNOX* genes (from *KNOTTED1* homeobox: the homeobox is the DNA sequence that encodes the homeodomain). *KNOX* genes are required for the initiation and maintenance of the shoot apical meristem (see *SHOOTMERISTEMLESS*: case studies 4.2 and 9.1). Consistent with this role, loss of *KN1* function causes a failure of shoot apical meristem development in the embryos of some maize lines.

The KN1 protein and *KN1* mRNA can be transported through plasmodesmata, hence *KN1* expression in one cell can alter transcription patterns in

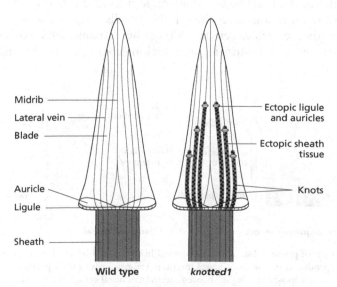

Midrib

Lateral vein

Blade

Auricle

Ligule

Sheath

Ectopic ligule and auricles

Ectopic sheath tissue

Knots

Wild type   ***knotted1***

Fig. 5.8 Wild-type and *knotted1* maize leaves. Tissue with sheath identity is coloured red. (Adapted from Freeling, 1992.)

neighbouring cells. In wild-type plants, *KN1* is expressed in the shoot apical meristem and in immature regions of the stem, but is silent at the sites of leaf initiation and in developing leaves. The dominant *kn1* mutations discussed above occur in the regulatory regions of *KN1* and cause ectopic synthesis of functional KN1 protein in the developing leaf. Specifically, mutant alleles are active in the tissue from which the lateral veins of the blade develop. The ectopic production of KN1 in a mutant leaf does not begin until the primordium is five plastochrons old. Hence the earlier boundary between the sheath and blade (established at P3, see above) remains mutable until at least this stage.

Why does *knotted1* affect the proximodistal axis?

The *kn1* phenotype is difficult to interpret because it is the consequence of a gain of function: a protein is present where it is normally absent. However, given the relationship between *KN1* expression and shoot apical meristem development, two models have been proposed to relate the *kn1* phenotype to proximodistal information in the leaf (Fig. 5.9). These models need not be mutually exclusive.

The proximodistal axis of the leaf may be patterned by a chemical signal produced by the shoot apical meristem. In the simplest case, the concentration gradient of a meristem-derived morphogen diffusing along the leaf would define axial position. Proximal regions of the leaf would experience a high concentration of the morphogen and adopt sheath fates. Distal regions would be exposed to a low concentration and adopt blade fates. Given that *KN1* expression is a feature of the shoot apical meristem, the KN1 protein may induce the production of such a morphogen. The ectopic production of KN1 in the mutant leaf blade could cause a local increase in the concentration of the morphogen, resulting in blade cells adopting sheath fates.

At the other extreme, cell fate along the proximodistal axis could be determined by a cell-intrinsic, temporal mechanism. As the maize leaf

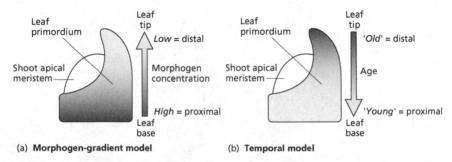

(a) **Morphogen-gradient model**   (b) **Temporal model**

Fig. 5.9 Models of proximodistal axis development in the maize leaf. (a) The concentration of a morphogen produced by the shoot apical meristem informs cells of their position in the leaf—proximal cells are exposed to a higher concentration than distal cells. (b) Cell fate is determined by the time elapsed since each region of the primordium adopted leaf identity—distal tissue is 'older' than proximal tissue.

primordium emerges from the apical meristem, additional meristem cells are recruited into the primordium base. Therefore, cells at the tip of the primordium may have acquired leaf identity earlier than cells at the primordium base. At the point at which the decision to adopt blade or sheath fates is made, cells might measure the length of time they have been part of the leaf. 'Older' leaf cells would adopt blade fates, 'younger' leaf cells would become sheath. Ectopic production of KN1 in the blade might reset the timing mechanism such that affected cells act as if they have been part of the leaf for a shorter period and so adopt sheath rather than blade fates.

Case study 5.3: Determinate nature of leaf development

A fundamental feature of angiosperm leaves is that they are determinate. This raises two questions. Firstly, what is the difference between the shoot apical meristem and the leaf primordium that makes the meristem indeterminate and the primordium determinate? Secondly, what governs the final size of the leaf, i.e. the length of the leaf axes? The answer to the first question may lie partly with the expression of *KNOX* genes in the apical meristem but not the leaf (see case study 5.2). There are few clues to help answer the second question. However, several lines of evidence suggest that mechanisms that regulate leaf size do so in accordance with the absolute dimensions of the leaf rather than with the number of leaf cells.

The *KNOX* genes and indeterminate development

As discussed in case study 5.2, in wild-type maize, *KN1* is expressed in the apical meristem but not at the sites of leaf initiation or in developing leaves. This observation can be extended to other *KNOX* genes, i.e. genes with a high level of sequence similarity to *KN1*, and to other species with simple leaves, e.g. tobacco and *Arabidopsis*. In all these cases, *KNOX* activity is lost from the sites of leaf initiation by the I1 stage and *KNOX* genes remain inactive during leaf development.

Loss-of-function mutations in *SHOOTMERISTEMLESS* (*STM*), an *Arabidopsis KNOX* gene, lead to the failure of meristem initiation during embryogenesis or, if the loss of function is partial, to premature meristem termination (see case studies 4.2 and 9.1). Hence *KNOX* genes are required to maintain the indeterminate state of the apical meristem. Furthermore, ectopic expression of *KNOX* genes in the leaf gives more 'indeterminate' patterns of development. For example, ectopic production of KN1 protein in *kn1* maize leaves causes extended cell divisions in the blade, producing the 'knots' (see case study 5.2). A more extreme phenotype is produced when a *KN1* construct is expressed at constitutively high levels in tobacco. This results in the development of very small leaves from which ectopic shoots are sometimes produced, indicating that high *KNOX* activity can induce the formation of shoot apical meristems on the leaf (Fig. 5.10).

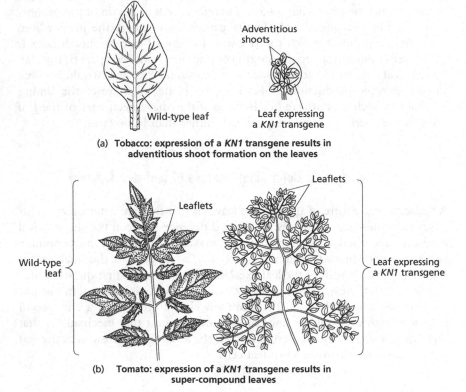

(a) **Tobacco: expression of a KN1 transgene results in adventitious shoot formation on the leaves**

(b) **Tomato: expression of a KN1 transgene results in super-compound leaves**

Fig. 5.10 The effects of expressing a *KN1* transgene in tobacco and tomato leaves. (a) High expression of a *KN1* transgene in tobacco results in a dramatic reduction in leaf size and the development of adventitious shoots on the leaves. (b) High expression of a *KN1* transgene in tomato results in the development of super-compound leaves. ((a) adapted from Sinha *et al.*, 1993; (b) adapted from Hareven *et al.*, 1996.)

These data suggest that absence of *KNOX* activity contributes to the determinate nature of leaf development. Interestingly, *KNOX* genes are expressed ectopically in the leaves of *phan* mutants of *Antirrhinum*, indicating that the *PHAN* gene acts to suppress *KNOX* activity in the leaf. As discussed in case study 5.1, *PHAN* encodes a MYB-like transcription factor and is expressed at the sites of leaf initiation and in leaf primordia. *PHAN* is homologous to the *ROUGH SHEATH2* (*RS2*) gene of maize and the *ASYMMETRIC LEAVES1* (*AS1*) gene of *Arabidopsis*. These two genes have similar expression patterns to *PHAN* and are also required to repress *KNOX* expression in the leaf. However, for reasons that are unclear, *phan*, *rs2* and *as1* mutants have different leaf phenotypes. Whereas the leaves of *phan* mutants have an abnormal or absent adaxial–abaxial axis (see case study 5.1), *rs2* mutants have leaves similar to those of *kn1* mutants (see case study 5.2), while *as1* mutants produce abnormal deeply lobed leaves from which shoots sometimes develop. The relationship between *AS1* and the *KNOX* genes is discussed further in Chapter 9.

Compound leaves

During the development of compound leaves, the leaf primordium branches one or more times before initiating leaflets (Chapter 1). This suggests that compound leaves follow a less determinate pattern of development than simple leaves. The mechanisms that control compound leaf development have been studied in tomato and pea. These species are only distantly related and appear to have evolved compound leaves independently. Reflecting this, the tomato leaf primordium initiates leaflets in a basipetal (tip-to-base) sequence, whereas the pea leaf primordium initiates leaflets and tendrils in an acropetal (base-to-tip) sequence. The two plants also appear to use different mechanisms to maintain the initial, branching phase of compound leaf development.

The *KN1* homologue of tomato is expressed in both the shoot apical meristem and the developing leaf. Furthermore, constitutive expression of a *KN1* transgene in tomato leads to 'super-compound' leaves in which primordium branching is reiterated many times (Fig. 5.10). This suggests that expression of *KNOX* genes in the tomato leaf delays the transition from indeterminate development, i.e. branching of the leaf primordium, to determinate development, i.e. the initiation of leaflets.

In contrast to tomato, *KNOX* gene expression is not detected in the pea leaf primordium. Instead, the early, indeterminate phase of pea leaf development is conferred by expression of the *UNIFOLIATA* (*UNI*) gene. *uni* mutants produce simple leaves, and also have abnormal flowers. *UNI* is a homologue of the *Arabidopsis* gene *LEAFY* (*LFY*) which functions to specify floral meristems (Chapter 9, and also see case study 5.4) In *Arabidopsis*, *LFY* is expressed in leaf and floral primordia; however, whereas loss of *LFY* function disrupts flower development, it has no obvious effect on *Arabidopsis* leaf development. It is possible, therefore, that the ancestral function of *UNI* is in the regulation of flower development, and that the function of *UNI* in compound leaf development arose secondarily.

Interestingly, loss of function of the tomato *LFY* homologue, *FALSIFLORA* (*FA*), also affects both leaf and flower development. The leaves of *fa* mutants produce fewer leaflets than wild type. This suggests that *FA* may also function in compound leaf development in tomato.

What determines the size of the leaf?

The final size and shape of a leaf depends on the position of the leaf on the shoot and on environmental conditions. However, given that isolated leaf primordia can develop into normal, albeit small, leaves in tissue culture, the leaf must contain an intrinsic mechanism for controlling its relative proportions. The nature of the mechanism is unknown but several lines of evidence suggest that it is linked to the absolute size of the leaf rather than the number and orientation of leaf cells.

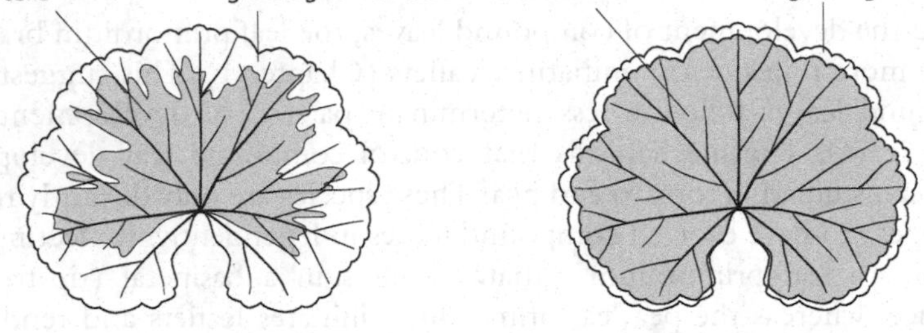

Green centre: mesophyll
consists of wild-type and
White1 cells

White edge: mesophyll
consists of only fast-
growing White1 cells

Green centre: mesophyll
consists of wild-type and
White2 cells

White edge: mesophyll
consists of only slow-
growing White2 cells

(a)　　**Green-White1-Green**

(b)　　**Green-White2-Green**

Fig. 5.11 Competition between fast- and slow-growing cells in chimeric *Pelargonium* leaves. The white border on each leaf represents the region in which the mesophyll is composed only of 'white' cells (see also Fig. 3.5). (a) White1 cells proliferate at the wild-type rate and contribute to a normal proportion of the leaf. (b) White2 cells proliferate slowly and contribute to a greatly reduced proportion of the leaf, however leaf shape and size are not affected. (Adapted from Stewart *et al.*, 1974.)

As discussed in Chapter 3, the patterns of cell divisions that generate the leaf are predictable but flexible. Leaf size does not, therefore, depend on strict lineage patterns that generate a fixed array of cells. The phenomenon of **competition** in the leaf provides further evidence for this (Fig. 5.11). In chimeric *Pelargonium* composed of wild-type cells and cells that are both chlorophyll-deficient and slow growing, the wild-type cells contribute to a far greater proportion of the leaf than expected from normal lineage patterns. This suggests that clones of leaf cells must compete to contribute to the leaf, with fast-growing clones contributing more than slow-growing clones. Despite the abnormal lineage pattern, the leaves of these plants are the normal size and shape.

Flexible lineage patterns could be accommodated by a mechanism that controlled the number of cells in the leaf. Such a mechanism may apply to the adaxial–abaxial axis of the leaf lamina in which there is a predictable array of tissue types: a typical dicot leaf lamina consists of adaxial epidermis, a palisade layer, a characteristic number of spongy mesophyll layers, and then abaxial epidermis. However, there is good evidence that growth along the proximo-distal and centrolateral axes is not related to cell numbers but rather to the absolute lengths of the axes.

Tobacco plants in which the rate of cell division is slowed by transgenic manipulation of the cell cycle possess fewer cells than wild-type plants. The transgenic plants are abnormally small as seedlings but eventually recover to an almost wild-type size. The later leaves on such plants contain many fewer cells than a wild-type leaf but are a very similar shape and size. The reduction in cell number is accommodated by a proportionate increase in cell size. The same phenomenon occurs almost throughout the transgenic plants, suggesting that size control generally is somewhat independent of cell number.

In a complementary experiment, transformation of tobacco plants with an inducible *AUXIN-BINDING PROTEIN1* (*ABP1*) gene allows an inducible increase in leaf cell expansion. Experimentally increased *ABP1* expression in transgenic tobacco results in leaf cells that are on average about twice the wild-type volume, presumably due to increased responsiveness to endogenous auxin. Despite the change in cell size, affected leaves develop to the wild-type size and shape. The developing leaf compensates for the increase in cell size by reducing cell proliferation.

An increase in size is also observed in polyploid cells. As discussed in case study 3.2, periclinal chimeras can be generated in which one layer of the shoot apical meristem consists of large polyploid cells, and the other layers contain normally sized diploid cells (see Fig. 3.6). Such chimeras develop normally. The larger size of the polyploid cells is compensated for by a reduction in the cell division rate of polyploid tissues. Hence when a polyploid leaf epidermis overlies diploid inner tissues, the rate of cell division in the epidermis is slower than that in the inner tissue, and the final morphology of the leaf is normal.

It is not known how the developing leaf can regulate its absolute size. However, the *angustifolia* (*an*) and *rotundifolia3* (*rot3*) mutants of *Arabidopsis* suggest the mechanisms are axis-specific. Leaves of both mutants have approximately the wild-type number of cells. Cells in *an* mutant leaves have reduced cell elongation specifically across the leaf width, producing narrow leaves of the wild-type length. In contrast, cells in *rot3* mutant leaves have reduced cell elongation specifically along the leaf length, producing short leaves of the wild-type width.

Interestingly, the regulation of growth in accordance with absolute dimensions rather than cell number is also observed in animals. In the fruit fly, for example, the number of cells in regions of the wing can be transgenically manipulated over a range of four-or fivefold without affecting the final wing size. The changes are accommodated by changes in cell size. As in the leaf, cells compete to contribute to the wing, although in this case, more slowly growing cells die. Furthermore, unlike a leaf primordium, the developing *Drosophila* wing is divided into compartments that cells will not cross between (described in Chapter 10), and competition in the wing is restricted to individual compartments.

Flowers

This section discusses the arrangement of floral organs relative to the axes of the flower. Case study 5.4 considers the radial axis: how does the concentric pattern of sepal, petal, stamen and carpel whorls develop? In species such as *Arabidopsis*, floral organs within each whorl share the same form. In many species, however, the flower also has an adaxial–abaxial axis (sometimes called a dorsoventral axis), for example the two petals at the top of the *Antirrhinum* flower (the adaxial petals) have a different form to the petal at the bottom of the flower (the abaxial petal). Case study 5.5 considers the adaxial–abaxial axis of the *Antirrhinum* flower.

Case study 5.4: Radial axis of the flower

Floral development in most angiosperms consists of variations of the same fundamental pattern. The floral meristem initiates an outer whorl of sepals, followed by a whorl of petals, a whorl of stamens, and finally a central whorl of carpels (Chapter 1). Genetic analysis, particularly in *Arabidopsis* and *Antirrhinum*, has identified many of the genes responsible for this pattern. For simplicity, this case study focuses on the development of the *Arabidopsis* flower. The flowers of *Arabidopsis* have four sepals, four petals, six stamens and two fused carpels (Fig. 5.12).

The ABC model

The mechanism that regulates organ identity along the radial axis of the flower was elucidated by the study of **floral homeotic mutants**. In these mutants, floral organs at one position in the flower are replaced by organs appropriate to another position. In some homeotic mutants, for example, petals are replaced by stamens. These mutations do not affect organ number or organ position, just organ identity. The existence of homeotic mutants therefore demonstrates that the regulation of organ formation in the flower is separable from the regulation of organ identity.

In most floral homeotic mutants, organ transformations affect pairs of adjacent whorls. Mutants of this type can be divided into three classes based on which pair of whorls is affected (Fig. 5.12). In **class A mutants**, the outer two whorls (whorls 1 and 2) are affected, with carpels replacing sepals and stamens replacing petals. In **class B mutants**, the middle two whorls (whorls 2 and 3) are affected, with sepals replacing petals and carpels replacing stamens. In **class C mutants**, the inner two whorls (whorls 3 and 4) are affected, with petals replacing stamens and a secondary flower replacing the carpels. The secondary flower of class C mutants repeats the developmental pattern of the primary flower, leading to the formation of an indeterminate series of concentric flowers: a pattern commonly seen in double varieties of garden flowers.

The above mutants define class A, class B and class C genes. The mutant phenotypes indicate that class A genes are required for the development of sepals and petals (whorls 1 and 2), class B genes are required for the development of petals and stamens (whorls 2 and 3) and class C genes are required for the development of stamens and carpels (whorls 3 and 4). This suggests that floral organ identity is controlled by a combinatorial system of gene activity, and is the basis of the **ABC model**.

This model proposes that class A, class B and class C genes are principally active in the pairs of whorls affected by their corresponding mutations. Sepals are specified by the activity of class A genes alone, petals are specified by the combination of class A and class B activity, stamens are specified by the combination of class B and class C activity, and carpels are specified by class C activity alone. To explain the homeotic phenotypes fully, two extra

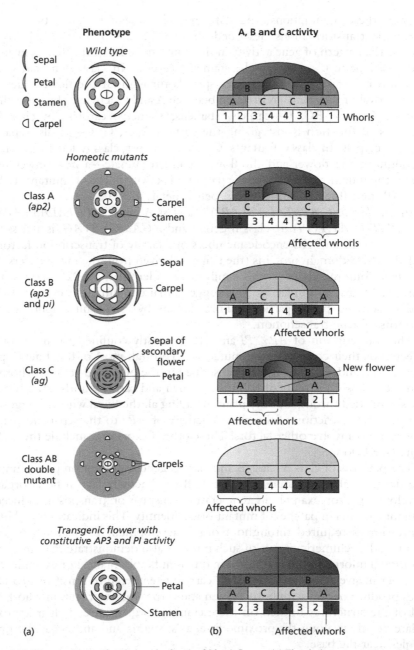

Fig. 5.12 Homeotic transformations in the *Arabidopsis* flower. (a) Floral phenotypes viewed from above; regions with altered organ identity are coloured red. (b) Patterns of class A, class B and class C gene activity viewed from the side and in cross-section. Whorls are numbered below the diagram and those in which the pattern of ABC gene activity differs from wild type are coloured dark red. (Adaped in part from Weigel & Meyerowitz, 1994; Parcy *et al.*, 1998.)

assumptions are required. Firstly, the model assumes that class A and class C genes mutually inhibit each other's activity. Secondly, it is assumed that class C function is required for determinate floral development.

With these assumptions, the ABC model successfully predicts all the homeotic transformations described above (Fig. 5.12). In the wild-type flower, the pattern of gene activity in the four whorls is A, AB, BC, C, giving the radial pattern sepals, petals, stamens, carpels. In class A mutants, A function is lost and class C activity spreads throughout the flower due to the removal of the mutual repression between A and C activity. This gives the pattern C, BC, BC, C—carpels, stamens, stamens, carpels. In class B mutants, B function is lost giving the pattern A, A, C, C—sepals, sepals, carpels, carpels. In class C mutants, C function is lost, class A activity spreads throughout the flower and the flower is indeterminate—a new flower replaces the fourth whorl. Hence the pattern of activity in class C mutants is A, AB, AB, new flower—giving sepals, petals, petals, new flower.

In *Arabidopsis*, *APETALA2* (*AP2*) is a class A gene, *APETALA3* (*AP3*) and *PISTILLATA* (*PI*) are class B genes, and *AGAMOUS* (*AG*) is a class C gene. *AP3*, *PI* and *AG* encode members of a family of transcription factors called MADS domain proteins (the name is derived from the initial letters of the first four members of the family to be identified). AP2 encodes an unrelated transcription factor. This suggests that combinations of class A, B and C activity specify particular floral organs by establishing appropriate patterns of gene transcription.

The transcription of *AP3*, *PI* and *AG* is mostly confined to the whorls affected by their corresponding mutations. Through most of floral development, *AP3* and *PI* (class B) are transcribed in the petal and stamen whorls, and *AG* (class C) is transcribed in the stamen and carpel whorls. *AP2* (class A) is expressed throughout the plant including all the floral whorls, suggesting that the restriction of the class A activity of *AP2* to the sepal and petal whorls must involve other factors. These other factors may include the *AP1* gene (see below).

The patterns of class A, class B and class C gene transcription persist until the flower is almost mature. Loss of A, B or C activity late in floral organ development, for example in temperature-sensitive or transposon-induced mutants, results in patches of mutant organ identity. This indicates that ABC expression is required throughout organ development to maintain organ identity. The chimeric organs of such mutants also demonstrate that internal positional information is conserved in different floral organs. For example, at non-permissive temperatures, plants carrying temperature-sensitive *ap3* alleles produce organs with both stamen and carpel characteristics in whorl 3. All of the structures formed on these chimeric organs arise at their appropriate positions along the proximodistal axis: stigma and anthers at the tip, ovules near the base.

The ABC model has been tested by generating double mutants and by driving ectopic expression of class A, class B or class C transgenes in the flower. For example, exactly as predicted by the model, class AB double mutants of *Arabidopsis* express *AG* in all whorls, giving the pattern C, C, C, C and flowers containing only carpel-like organs (Fig. 5.12). Also as predicted, ectopic expression of *AP3* and *PI* throughout the flower produces the pattern AB, AB, BC, BC and the phenotype petals, petals, stamens, stamens. Interestingly, triple mutants that lack class A, B and C activity produce flowers in which all whorls contain leaf-like organs. This supports the hypothesis that floral organs are modified leaves.

The *SEPALLATA* genes

The effects of the loss of *SEPALLATA* (*SEP*) gene function in *Arabidopsis* suggest that additional transcription factors are required to specify petals, stamens and carpels. There are three *SEP* genes, *SEP1*, *SEP2* and *SEP3*, all of which encode closely related MADS domain transcription factors. Single *sep1*, *sep2* and *sep3* mutants have almost wild-type phenotypes but the *sep1 sep2 sep3* triple mutant has nested flowers consisting only of sepals, i.e. a whorled pattern of sepal, sepal, sepal, new flower. This is a **phenocopy** of the class B class C double mutant, which has only class A expression in all four whorls, and suggests that the *SEP* genes are required to allow class B and class C activity. Consistent with this role, expression of the *SEP* genes is largely confined to the inner three whorls of the flower.

As discussed above, the *SEP* genes, the *AP3* and *PI* genes (class B), and the *AG* gene (class C) all encode MADS domain transcription factors. The AP3 and PI proteins form a heterodimer, which can interact with the SEP3 and AG proteins. Similarly, the AG protein can interact with the SEP1, SEP2 and SEP3 proteins. This suggests that MADS domain proteins act in multimeric complexes.

Establishing the pattern of A, B and C activity

The ABC model successfully explains how the identity of each type of floral organ is specified: by the combinatorial action of transcriptional regulators. This begs the question of how the pattern of class A, class B and class C activity is established. The mechanism that establishes class B activity is best understood and is considered first.

Establishing the pattern of class B activity

The transcription of class B genes in the flower is partly established by the genes that function to specify floral meristem identity. The transition from vegetative to reproductive development involves the coordination of several endogenous and environmental factors. Central to the transition to flowering is an increase in expression of the *LFY* gene at the sites of floral meristem formation, followed by the transcription of *AP1* and *CAULI-FLOWER* (*CAL*) in the developing floral meristem (discussed in Chapter 9). All of these genes encode transcription factors (*AP1* and *CAL* encode closely related MADS box genes).

The effects of *LFY* and *AP1* on the transcription of class B genes in the flower have been studied in detail. For example, *AP3* transcription is greatly reduced in the flowers of *lfy* mutants, and virtually absent in the lateral organs of *lfy ap1* double mutants. This suggests that *LFY* and *AP1* are required to establish *AP3* transcription in the flower. *LFY* and *AP1* are both

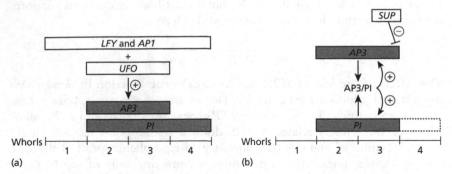

Fig. 5.13 Establishing the pattern of class B gene expression. (a) *LFY*, *AP1* and *UFO* promote the transcription of the class B genes *AP3* and *PI* in the floral meristem. *AP3* is expressed in whorls 2 and 3, reflecting the region of *UFO* expression. It is unclear why *PI* is initially expressed in whorls 2, 3 *and* 4. (b) Later in meristem development, the expression of *AP3* and *PI* is maintained through the action of the AP3–PI heterodimer. Since the heterodimer is not present in whorl 4, *PI* expression ceases in the centre of the flower. *SUP* is required to prevent *AP3* expression from spreading into whorl 4. Arrows accompanied by a 'plus' sign indicate positive regulation. Lines ending in a bar and accompanied by a 'minus' sign indicate negative regulation.

expressed uniformly across the floral meristem at the time *AP3* is first transcribed, whereas *AP3* transcription is always restricted to whorls 2 and 3. The regional pattern of *AP3* transcription appears to depend on the *UNUSUAL FLORAL ORGANS (UFO)* gene (Fig. 5.13). *UFO* is initially expressed throughout the floral primordium but its expression is rapidly restricted to whorls 2 and 3. Loss of *UFO* function causes variable, pleiotropic abnormalities in both floral initiation and development. A common phenotype in *ufo* flowers is the absence of normal petals and stamens, correlated with a reduction in *AP3* and *PI* activity. Studies on embryo and seedling development show that *UFO* is expressed in a 'class B' position in vegetative shoot apical meristems. *UFO* therefore links the radial pattern of the floral meristem to radial information in the vegetative meristem. *UFO* encodes a protein involved in the regulation of protein degradation. One possibility is that *UFO* expression results in the destruction of proteins that negatively regulate *AP3* transcription.

Analysis of *PI* gene expression indicates that its transcription is also promoted by *LFY*, *AP1* and *UFO*. However, unlike *AP3*, *PI* is initially transcribed in whorls 2, 3 and 4 of the floral meristem and is only later restricted to whorls 2 and 3 (Fig. 5.13). The mechanism that silences *PI* transcription in whorl 4 is probably based on mutual dependency between *PI* and *AP3* expression. As described above, the AP3 and PI proteins bind to each other to form a heterodimeric transcription factor which may also interact with other MADS domain proteins. The targets of the AP3–PI heterodimer include the *AP3* and *PI* genes themselves, and the heterodimer is required for maintenance of class B gene transcription. This maintenance function is essential because *LFY*, which establishes both *AP3* and *PI* expression, is only expressed at the early stages of floral development. Consequently, although the transcription of *AP3* and *PI* is established independently, loss of function of either gene leads to the cessation of transcription from both. Since *AP3* is not expressed in whorl 4, and hence the AP3–PI heterodimer is absent at that position, the initial transcription of *PI* in whorl 4 is not maintained (Fig. 5.13b).

The exclusion of *AP3* expression from whorl 4 requires the *SUPERMAN* (*SUP*) gene (Fig. 5.13b). In floral meristems of *sup* mutants, *AP3* is initially expressed in whorls 2 and 3 (as in wild type) but *AP3* expression subsequently spreads into whorl 4. Consequently, both *AP3* and *PI* are transcribed in the centre of *sup* mutant flowers, giving a combination of class B and class C expression and the conversion of carpels to stamens. *SUP* encodes a putative transcription factor and is expressed in a narrow band of whorl 3 cells next to the border with whorl 4 (Fig. 5.13b). Therefore, *SUP* may act non-cell-autonomously to inhibit *AP3* expression in whorl 4 cells.

Establishing the pattern of class A and class C activity

Central to the ABC model is the mutual inhibition of class A and class C function. This suggests that class A and class C genes are involved in establishing each other's patterns of activity. As yet, however, the mechanism that determines the patterns of class A and class C activity is poorly understood.

As discussed earlier, *AP2* is expressed in all four whorls of the flower and therefore the restriction of the class A activity of *AP2* to whorls 1 and 2 requires another factor. A candidate is the floral meristem identity gene *AP1*. *AP1* is initially expressed throughout the floral meristem (see above), but its expression is later restricted to whorls 1 and 2. This restriction is dependent on the class C gene *AG* (Fig. 5.14), since *AP1* transcription persists across the whole flower in *ag* mutants. Loss of *AP1* function delays the transition to flowering. It also results in a transformation of sepals to bract-like organs, and the absence of petals. Therefore, as well as its role in specifying floral development, *AP1* may act in concert with *AP2* to specify sepal and petal identity. Consequently, *AP1* is sometimes considered to be a class A gene.

What, then, establishes the pattern of *AG* expression? Consistent with its class C function, *AG* transcription is limited to whorls 3 and 4, but to date most of the factors known to regulate *AG* expression are uniformly distributed across the flower (Fig. 5.14). *AG* transcription is delayed in the flowers of *lfy* mutants, indicating that *LFY* promotes *AG* expression. *AG* is transcribed throughout the flower in *ap2* mutants, consistent with the mutual inhibition of class A and class C activity and indicating that *AP2* inhibits *AG* expression in whorls 1 and 2. *AG* is also expressed throughout the flower in *leunig* (*lug*) mutants, resulting in class A homeotic transformation of sepals to carpels and petals to stamens (or their absence). This indicates that the *LUG* gene is required to restrict *AG* transcription to whorls 3 and 4. Consistent with this role, *LUG* encodes a putative transcriptional co-repressor, a protein that acts with a transcription factor to repress transcription. Despite the class A transformations in *lug* single mutants, *lug ag* double mutants develop normal sepals and petals, indicating that *LUG* is only required to restrict *AG* transcription and not to specify sepal and petal identity.

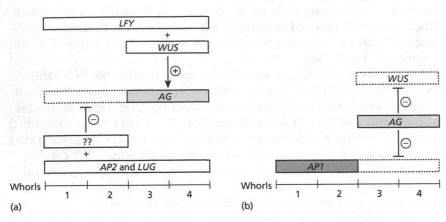

Fig. 5.14 Establishing the patterns of class A and class C activity. (a) The transcription of the class C gene *AG* is promoted by *LFY* and *WUS*, and is inhibited by *AP2* and *LUG*. A putative asymmetrically distributed inhibitor may act with *AP2* and *LUG* to prevent *AG* expression in whorls 1 and 2. (b) *AP1* is initially transcribed across the floral meristem but its expression is inhibited by *AG*. Consequently, once *AG* expression is initiated, *AP1* expression is restricted to whorls 1 and 2, in which *AP1* contributes to class A function. *AG* also inhibits *WUS* expression. Since *WUS* is required to maintain meristematic activity (see text), this results in determinate floral development. Arrows accompanied by a 'plus' sign indicate positive regulation. Lines ending in a bar and accompanied by a 'minus' sign indicate negative regulation.

LFY, *AP2* and *LUG* are expressed across the whole flower. Therefore, the restriction of *AG* expression to whorls 3 and 4 must depend on additional factors that are distributed asymmetrically along the radial axis of the flower (Fig. 5.14a). In the shoot apical meristem, the radial axis is patterned by the *WUSCHEL* (*WUS*) gene, whose expression is restricted to the centre of the meristem by the action of the *CLAVATA* (*CLV*) genes (see case study 9.1). Since the floral meristem represents a modified shoot apical meristem, it is not surprising that *WUS* also plays a role in ABC patterning. As in the shoot apical meristem, *WUS* is expressed in the centre of the floral meristem and is required to maintain meristem activity. *WUS* encodes a homeodomain transcription factor that binds to regulatory sequences in the second intron of the *AG* gene. The LFY transcription factor also binds to this region of the *AG* gene and, together, WUS and LFY activate *AG* expression in the centre of the flower.

As described above, *ag* flowers are indeterminate, producing a nested pattern of sepals and petals (Fig. 5.12). Thus, in addition to specifying stamen and carpel identity, *AG* is required for the determinate development of the floral meristem. In wild-type flowers, *WUS* expression ceases as the carpels are initiated. In contrast, *WUS* is continuously active in the nested flowers of *ag* mutants. Furthermore, the flowers of *ag wus* double mutants terminate early, like those of *wus* single mutants, demonstrating that *WUS* is required for the indeterminate development of *ag* flowers. These data suggest that *AG* acts to inhibit *WUS* in the centre of the floral meristem (Fig. 5.14b). Therefore, *WUS* and *AG* are components of a negative feedback loop: *WUS* activates *AG*, establishing the radial pattern of floral organs; later, *AG* inhibits *WUS*, making the floral meristem determinate.

Evolution of the ABC mechanism

In evolutionary terms, stamens and carpels are the fundamental units of the flower. Consistent with this, the specification of carpels by class C gene expression and of stamens by a combination of class B and class C expression appears to be universal in the angiosperms. Furthermore, it is possible that class B and class C genes act in other groups, such as the gymnosperms, to specify and distinguish microspore production from megaspore production (Chapter 1). In contrast to stamens and carpels, sepals and petals are thought to have evolved independently in several different groups of angiosperms. For example, in most dicots, including *Arabidopsis*, petals evolved from modified stamens, whereas in some primitive dicots, such as the Magnoliales, petals appear to have arisen from modified bracts.

Consistent with more recent evolution of petals and sepals compared to stamens and carpels, class A activity in *Arabidopsis* arises in a very different manner to class B and class C activity. As discussed above, class B and C activity is provided by closely related MADS box genes that are expressed in the appropriate pairs of whorls in the developing flower. In contrast, class A activity is provided by *AP2*, which is ubiquitously expressed and encodes an unrelated transcription factor, and by *AP1*, which is a MADS box gene but which also functions to specify floral meristems. An attractive hypothesis is that class A activity is not truly separable from the regulation of floral meristem identity. In this model, sepal identity occurs by default in floral organs that do not express either class B or class C genes, while class B expression in the absence of class C activity specifies petal identity, consistent with the evolution of petals from modified stamens. The class A phenotype of *ap2* mutants can be explained by the ectopic expression of *AG* (class C) in whorls 1 and 2, while the *ap1* floral phenotype (sepals replaced by bracts, and petals absent) can be considered as a partial failure to specify floral meristem identity.

In this context, it is interesting to note that no unambiguous class A gene has been identified in *Antirrhinum*. Class A mutants of *Antirrhinum* all result from mutations in the regulatory regions of the *Antirrhinum* class C gene, *PLENA*. These mutations cause ectopic *PLENA* expression in whorls 1 and 2.

Other aspects of the radial axis

There are other characteristics of the radial axis of the flower besides organ identity. One of these is the initial shape of the floral organ primordia in each whorl. Clonal analysis of wild-type flowers and homeotic mutants suggests that the shape of floral organ primordia is determined separately from the identity of the floral organs.

The initial shapes of the floral organ primordia of *Arabidopsis* were investigated using **sector boundary analysis**. Clones of marked cells were generated prior to flower development and the positions of the boundaries between marked and unmarked sectors on the floral organs were analysed. The number of positions at which boundaries occur indicates how many cells

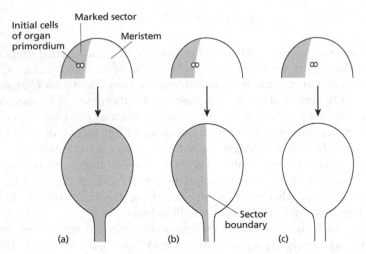

Fig. 5.15 Sector boundary analysis. The diagram shows a sector boundary analysis of an organ initiated as a pair of cells aligned perpendicular to the boundary between marked and wild-type clones at the shoot apex. In this case, marked sectors will encompass: (a) the whole organ, (b) half of the organ or (c) none of the organ.

initiate an organ. The orientations of the boundaries indicate the shape of the initiating group. For example, if sector boundaries only ever occur along the centre of an organ, and so divide the organ in half, that organ is probably initiated by only two adjacent cells (Fig. 5.15).

Sector boundary analysis indicates that sepals and carpels are initiated by a line of eight cells, petals by a pair of cells, and stamens by four cells in a block. In *ap3* and *pi* mutants, petals are replaced by sepals. However, sector boundary analysis of mutant flowers shows that the second whorl sepals of the mutants are initiated by a group of two cells, like the petals that they replace. Hence the initial shapes of organ primordia appear to be determined by factors other than the expression of the ABC genes.

Case study 5.5: Adaxial–abaxial axis of the *Antirrhinum* flower

In *Antirrhinum*, flowers project horizontally from a vertical inflorescence and floral development is organized around two axes: the **radial axis**, leading from the sepals to the carpels; and the **adaxial–abaxial axis** (sometimes called the dorsoventral axis) leading from the top of the flower to the bottom. This case study considers the development of the adaxial–abaxial axis, discussing the genetic control of development along the axis and the initial establishment of adaxial–abaxial asymmetry.

The *Antirrhinum* flower

Antirrhinum floral meristems are initiated in the axils of bract primordia, which are generated by the inflorescence meristem. The **adaxial** side of the floral meristem, that facing towards the longitudinal axis of the inflorescence, forms the top of the flower. The **abaxial** side of the floral meristem, that facing away from the longitudinal axis, forms the bottom of the flower.

The mature flower has an outer whorl of five sepals. There is a second whorl of five petals which are fused along part of their length to produce the **corolla tube**. The third whorl contains four complete stamens and a single, rudimentary stamen called the **staminode**. Finally, the fourth whorl consists of two, fused carpels (Fig. 5.16).

The adaxial–abaxial axis of the flower is most clearly expressed in the petal and stamen whorls (Fig. 5.16). The five petals are arranged such that there are two adaxial petals, two lateral petals and a single abaxial petal. The plane of symmetry passes between the two adaxial petals and bisects the abaxial petal. Each class of petal has its own distinctive appearance. Furthermore, adaxial and lateral petals are internally asymmetric along the adaxial–abaxial axis of the flower.

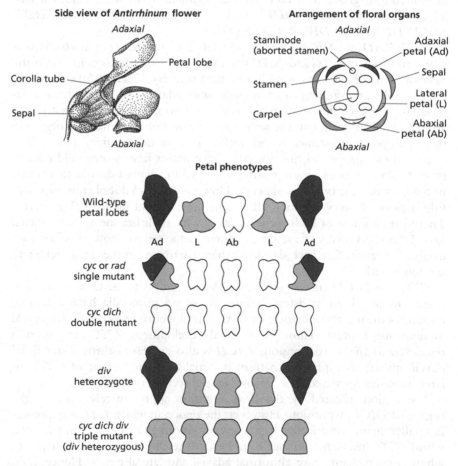

Fig. 5.16 Petal phenotypes of *Antirrhinum* mutants with altered floral adaxial–abaxial polarity. Note that in the *div* single mutant and the *cyc dich div* triple mutant, the *div* mutation is heterozygous. The *div* homozygote has a more severe phenotype (see text). (Adapted in part from Coen, 1996.)

Each of the five stamens is flanked on the outside by a pair of petals. As a consequence, there is a single adaxial stamen, two lateral stamens, and two abaxial stamens. The adaxial stamen aborts to produce the staminode. The remaining stamens have distinct lateral or abaxial forms. Lateral stamens are shorter than abaxial stamens. In addition, the filaments of all four complete stamens twist so that the anthers face abaxially—an effect of the adaxial–abaxial axis on the symmetry of individual stamens.

Genetic control of adaxial–abaxial asymmetry

Mutations that transform bilaterally symmetric flowers into flowers with radial symmetry are called **peloric**, from the Greek for 'monster'. Mutations that cause a partial disruption to the adaxial–abaxial axis are called **semi-peloric**. The study of peloric and semipeloric mutations in *Antirrhinum* has identified four genes involved in the development of the adaxial–abaxial axis. These genes are: *CYCLOIDEA* (*CYC*), *RADIALIS* (*RAD*), *DICHOT-OMA* (*DICH*) and *DIVARICATA* (*DIV*).

CYC, *RAD* and *DICH* promote adaxial characteristics (Fig. 5.16). Recessive mutations in *CYC* and *RAD* produce semipeloric flowers in which the development of the adaxial region is most severely affected. Adaxial petals, for example, are often replaced by petals with adaxial and lateral characteristics. The shape of the lateral petals is also abnormal, resembling that of the abaxial petal. Interestingly, the semipeloric flowers often contain six organs in the sepal, petal and stamen whorls rather than the normal five.

In contrast to *cyc* and *rad* flowers, *dich* mutants have a very mild phenotype. Mutant flowers appear relatively normal but have a deeper than usual notch between the two adaxial petals. However, *cyc dich* double mutants have fully peloric flowers in which all organs have abaxial identity (Fig. 5.16). Therefore, the loss of both *cyc* and *dich* activity abolishes the adaxial–abaxial axis. Like the flowers of *cyc* single mutants, the peloric flowers of *cyc dich* double mutants often have six rather than five organs in the sepal, petal and stamen whorls.

CYC and *DICH* encode similar DNA-binding proteins. *CYC* expression begins in the floral meristem before any organ primordia have emerged, continues during the initiation of all the floral whorls, and is still detectable in maturing flowers. Throughout floral development, *CYC* expression is restricted to the adaxial region. *DICH* is also expressed throughout floral development; its expression pattern is initially similar to that of *CYC* but later becomes restricted to a narrow, adaxial section of the flower.

The region affected by the *dich* mutation approximately matches the region of *DICH* expression. However, the region in which *CYC* is expressed is smaller than that affected by the *cyc* mutation. For example, in the petal whorl, *CYC* transcription can only be detected in the two adaxial petals, whereas *cyc* mutants have abnormal adaxial and lateral petals. Hence *CYC* activity in the adaxial section of the petal whorl influences the development of lateral petals, indicating intercellular signalling between the adaxial and lateral sections of the flower. Similarly, the radially symmetric flowers of *cyc dich* double mutants suggest that signalling between the region of *CYC* and *DICH* expression and the rest of the floral meristem occurs as the adaxial–abaxial axis is established.

The conclusion that *CYC* functions to promote adaxial organ identity is confirmed by the *backpetals* mutant, in which a transposon inserted in the regulatory region of the *CYC* gene causes ectopic *CYC* expression in the lateral and abaxial regions of the flower. The lateral and abaxial petals of *backpetals* flowers have adaxial characteristics.

The semidominant *div* mutation produces a semipeloric flower but the mutation affects the abaxial rather than the adaxial region (Fig. 5.16). For example, the abaxial petal of *div* heterozygotes has lateral petal characteristics. In homozygous *div* mutants, the phenotype of the abaxial petal is even more lateralized, and lateral petals are also affected such that the abaxial half of each lateral petal now resembles the adaxial half.

Recessive mutations usually result from a loss of function and are therefore relatively easy to interpret. Dominant and semidominant mutations can be the consequence of either a loss or a gain of function and are more difficult to interpret. In the case of *div*, experiments varying the dosage of the wild-type and mutant alleles indicate that the mutation results in a loss of function. The appearance of *div* flowers therefore suggests that *DIV* is required for the development of abaxial characteristics. Given that *cyc*, *dich* and *rad* mutants all have abnormal adaxial organ identity, a likely model is that the development of the adaxial–abaxial axis involves the imposition of both adaxial and abaxial organ identities on a default lateral form. Consistent with this model, *cyc dich div* triple mutants have radially symmetric flowers in which all organs adopt lateral fates (Fig. 5.16).

Basis of adaxial–abaxial asymmetry

The effects of mutations in the *CENTRORADIALIS* (*CEN*) gene demonstrate that the adaxial–abaxial asymmetry of the *Antirrhinum* flower requires positional information from the inflorescence apex (Fig. 5.17). Wild-type *Antirrhinum* flowers develop in the axils of bracts on an indeterminate inflorescence. *cen* mutants have a determinate inflorescence that ends in a terminal flower. The axillary flowers on a mutant inflorescence have normal adaxial–abaxial asymmetry, but the terminal flower is radially symmetric. This suggests that adaxial–abaxial asymmetry depends on the asymmetric environment of the axillary flower, i.e. a position between the bract primordium and the centre of the inflorescence meristem.

All petals on the radially symmetric terminal flower of *cen* mutants resemble the wild-type abaxial petal. As described above, *CYC*, *RAD* and *DICH* promote adaxial floral organ identity, whereas *DIV* promotes abaxial identity. Therefore, the presence of only abaxial-type petals on the *cen* terminal flower suggests that *DIV* functions in this context but that *CYC*, *RAD* and *DICH* do not.

The flowers of *cyc dich* double mutants are radially symmetric despite developing in the normal, axillary position. This suggests that *CYC* and

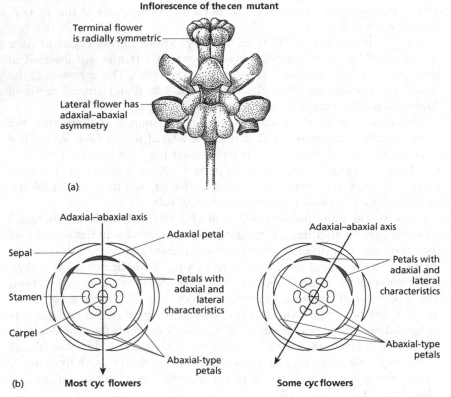

Fig. 5.17 Establishing the adaxial–abaxial axis of the *Antirrhinum* flower. (a) The *cen* mutant produces a terminal flower with radial symmetry. This suggests that the adaxial–abaxial axis of wild-type *Antirrhinum* flowers depends on their lateral position on the inflorescence. (b) *cyc* flowers have a variable relationship between the adaxial–abaxial axis and the arrangement of floral organs. This suggests that *CYC* is required for the consistent orientation of the adaxial–abaxial axis in the wild-type flower. ((a) adapted from Bradley *et al.*, 1996; (b) adapted from Luo *et al.*, 1996.)

DICH are required for the mechanism that interprets the asymmetric position of the floral primordium to generate the adaxial–abaxial axis. Observation of semipeloric *cyc* flowers with six petals shows that *CYC* is also required for the consistent alignment of the adaxial–abaxial axis relative to the arrangement of floral organs (Fig. 5.17). In most six petalled *cyc* flowers, the adaxial–abaxial axis bisects the uppermost petal. In some of the flowers, however, the adaxial–abaxial axis passes between two of the upper petals.

Conclusions

The primary radial and longitudinal axes of the plant are established during embryogenesis. During shoot development, these primary axes are interpreted to direct axial development in leaves. Of the three leaf axes—proximodistal, adaxial–abaxial and centrolateral—only the development of the adaxial–

abaxial axis is understood in any detail. The leaf primordium is polarized along the adaxial–abaxial axis by a signal from the shoot apical meristem. Axis maintenance and elaboration then become intrinsic to the primordium. The adaxial–abaxial axis appears necessary for the growth of the leaf lamina and also for the maintenance of the shoot apical meristem, demonstrating a reciprocal relationship between meristem and leaf.

It is probable that the mechanisms regulating axial development in the leaf also act in floral organs, since these evolved from leaves. The identity of a floral organ depends on its position along the radial axis of the flower, which is established *de novo* in each floral meristem. In addition, in plants with bilaterally symmetric flowers, floral organ development is regulated by position along the adaxial–abaxial axis of the flower, which is derived from axial information in the primary shoot. Positional information along the radial axis of the floral meristem is at least partly homologous to positional information in the vegetative shoot apical meristem. However, the ways in which positional information in the flower is decoded to produce patterns of floral organs vary between species. This reflects the independent evolution of sepals and petals, and also of bilateral symmetry, in flowers of different groups of angiosperms.

Further reading

General

Leaves

Bowman, J.L. (2000) Axial patterning in leaves and other lateral organs. *Current Opinion in Genetics and Development* **10**, 399–404.

Freeling, M. (1992) A conceptual framework for maize leaf development. *Developmental Biology* **153**, 44–58.

Smith, L.G. & Hake, S. (1992) The initiation and determination of leaves. *Plant Cell* **4**, 1017–1027.

Sylvester, A.W., Smith, L. & Freeling, M. (1996) Aquisition of identity in the developing leaf. *Annual Review of Cell and Developmental Biology* **12**, 257–304.

Flowers

Coen, E.S. (1991) The role of homeotic genes in flower development and evolution. *Annual Review of Plant Physiology and Plant Molecular Biology* **42**, 241–279.

Coen, E.S. (1996) Floral symmetry. *EMBO Journal* **15**, 6777–6788.

Case study 5.1: Adaxial–abaxial axis of the leaf

PHANTASTICA

Waites, R. & Hudson, A. (1995) *phantastica*: a gene required for dorsoventrality of leaves in *Antirrhinum majus*. *Development* **121**, 2143–2154.

Waites, R., Selvadurai, H.R.N., Oliver, I.R. & Hudson, A. (1998) The *PHANTASTICA* gene encodes a MYB transcription factor involved in growth and dorsoventrality of lateral organs in *Antirrhinum*. *Cell* **93**, 779–789.

PINHEAD and ARGONAUTE1

Lynn, K., Fernandez, A., Aida, M. *et al.* (1999) The *PINHEAD/ZWILLE* gene acts pleiotro-
pically in *Arabidopsis* development and has overlapping functions with the *ARGONAUTE1*
gene. *Development* **126**, 113.

PHABULOSA

McConnell, J.R. & Barton, M.K. (1998) Leaf polarity and meristem formation in *Arabidopsis*.
Development **125**, 2935–2942.
McConnell, J.R., Emery, J., Eshed, Y. *et al.* (2001) Role of *PHABULOSA* and *PHAVOLUTA* in
determining radial patterning in shoots. *Nature* **411**, 709–713.

The YABBY family

Siegfried, K.R., Eshed, Y., Baum, S.F. *et al.* (1999) Members of the *YABBY* gene family specify
abaxial cell fate in *Arabidopsis*. *Development* **126**, 4117–4128.

LAM1

McHale, N.A. (1992) A nuclear mutation blocking initiation of the lamina in leaves of *Nicotiana
sylvestris*. *Planta* **186**, 355–360.
McHale, N.A. & Marcotrigiano, M. (1998) *LAM1* is required for dosoventrality and lateral
growth of the leaf blade in *Nicotiana*. *Development* **125**, 4235–4243.

LEAFBLADELESS1

Timmermans, M.C.P., Schultes, N.P., Jankovsky, J.P. & Nelson, T. (1998) *Leafbladeless1* is
required for dosoventrality of lateral organs in maize. *Development* **125**, 2813–2823.

Case study 5.2: Proximodistal axis of the leaf

KNOTTED1

Freeling, M. & Hake, S. (1985) Developmental genetics of mutants that specify *Knotted* leaves
in maize. *Genetics* **111**, 617–634.
Jackson, D., Veit, B. & Hake, S. (1994) Expression of maize *KNOTTED-1* related homeobox
genes in the shoot apical meristem predicts patterns of morphogenesis in the vegetative shoot.
Development **120**, 405–413.
Smith, L., Greene, B., Veit, B. & Hake, S. (1992) A dominant mutation in the maize homeobox
gene, *Knotted-1*, causes its ectopic expression in leaf cells with altered fates. *Development* **116**,
21–30.
Vollbrecht, E., Reiser, L. & Hake, S. (2000) Shoot meristem size is dependent on inbred
background and presence of the maize homeobox gene, *knotted1*. *Development* **127**,
3161–3172.

Models of axis development

Freeling, M. (1992) A conceptual framework for maize leaf development. *Developmental Biology*
153, 44–58.
Hake, S., Char, B.R., Chuck, G. *et al.* (1995) Homeobox genes in the functioning of plant
meristems. *Philosophical Transactions of the Royal Society of London, Series B* **350**, 45–51.

Case study 5.3: Determinate nature of leaf development

The KNOX genes

Hake, S., Char, B.R., Chuck, G. *et al.* (1995) Homeobox genes in the functioning of plant meristems. *Philosophical Transactions of the Royal Society of London, Series B* **350**, 45–51.

The PHAN homologues

Byrne, M.E., Barley, R., Curtis, M. *et al.* (2000) *Asymmetric leaves1* mediates leaf patterning and stem cell function in *Arabidopsis*. *Nature* **408**, 967–971.

Timmermans, M.C., Hudson, A., Becraft, P.W. & Nelson, T. (1999) ROUGH SHEATH2: a Myb protein that represses *knox* homeobox genes in maize lateral organ primordia. *Science* **284**, 151–153.

Tsiantis, M., Schneeberger, R., Golz, J.F., Freeling, M. & Langdale, J.A. (1999) The maize *rough sheath2* gene and leaf development programs in monocot and dicot plants. *Science* **284**, 154–156.

Waites, R. & Hudson, A. (1995) *phantastica*: a gene required for dorsoventrality of leaves in *Antirrhinum majus*. *Development* **121**, 2143–2154.

Waites, R., Selvadurai, H.R.N., Oliver, I.R. & Hudson, A. (1998) The *PHANTASTICA* gene encodes a MYB transcription factor involved in growth and dorsoventrality of lateral organs in *Antirrhinum*. *Cell* **93**, 779–789.

Tomato leaf development

Hareven, D., Gutfinger, T., Parnis, A., Eshed, Y. & Lifschitz, E. (1996) The making of a compound leaf: genetic manipulation of leaf architecture in tomato. *Cell* **84**, 735–744.

Pea leaf development

Gourlay, C.W., Hofer, J.M.I. & Ellis, T.H.N. (2000) Pea compound leaf architecture is regulated by interactions among the genes *UNIFOLIATA, COCHLEATA, AFILA,* and *TENDRIL-LESS*. *Plant Cell* **12**, 1279–1294.

Hofer, J., Turner, L., Hellens, R. *et al.* (1997) *UNIFOLIATA* regulates leaf and flower morphogenesis in pea. *Current Biology* **7**, 581–587.

Regulation of leaf dimensions

Hemerly, A., de Almeida Engler, J., Bergounioux, C. *et al.* (1995) Dominant negative mutants of the Cdc2 kinase uncouple cell division from iterative plant development. *EMBO Journal* **14**, 3925–3936.

Jones, A.M., Im, K-H., Savka, M.A. *et al.* (1998) Auxin-dependent cell expansion mediated by overexpressed auxin-binding protein 1. *Science* **282**, 1114–1117.

Stewart, R.N., Semeniuk, P. & Dermen, H. (1974) Competition and accommodation between apical layers and their derivatives in the ontogeny of chimeral shoots of *Pelargonium X hortorum*. *American Journal of Botany* **61**, 54–67.

Tsuge, T., Tsukaya, H. & Uchimiya, H. (1996) Two independent and polarized processes of cell elongation regulate leaf blade expansion in *Arabidopsis thaliana* (L.) Heynh. *Development* **122**, 1589–1600.

Size control during animal development

Conlon, I. & Raff, M. (1999) Size control in animal development. *Cell* **96**, 235–244.

Day, S.J. & Lawrence, P.A. (2000) Measuring dimensions: the regulation of size and shape. *Development* **127**, 2977–2989.

Case study 5.4: Radial axis of the flower

The ABC model and interactions between organ identity genes

Bowman, J.L., Smyth, D.R. & Meyerowitz, E.M. (1991) Genetic interactions among floral homeotic genes of *Arabidopsis*. *Development* **112**, 1–20.
Carpenter, R. & Coen, E.S. (1990) Floral homeotic mutations produced by transposon-mutagenesis in *Antirrhinum majus*. *Genes and Development* **4**, 1483–1493.
Irish, V.F. (1999) Patterning the flower. *Developmental Biology* **209**, 211–220.
Weigel, D. & Meyerowitz, E.M. (1994) The ABCs of floral homeotic genes. *Cell* **78**, 203–209.

The SEPALLATA genes

Gutierrez-Cortines, M.E. & Davies, B. (2000) Beyond the ABCs: ternary complex formation in the control of floral organ identity. *Trends in Plant Science* **5**, 471–476.
Pelaz, S., Ditta, G.S., Baumann, E., Wisman, E. & Yanofsky, M.F. (2000) B and C floral organ identity functions require *SEPALLATA* MADS-box genes. *Nature* **405**, 200–203.

Sector boundary analysis of floral organ initiation

Bossinger, G. & Smyth, D.R. (1996) Initiation patterns of flower and floral organ development in *Arabidopsis thaliana*. *Development* **122**, 1093–1102.

Establishing the pattern of organ identity gene expression

Busch, M.A, Bomblies, K. & Weigel, D. (1999) Activation of a floral homeotic gene in *Arabidopsis*. *Science* **285**, 585–587.
Lenhard, M., Bohnert, A., Jürgens, G. & Laux, T. (2001) Termination of stem cell maintenance in *Arabidopsis* floral meristems by interactions between *WUSCHEL* and *AGAMOUS*. *Cell* **105**, 805–814.
Lohmann, J.U., Hong, R.L., Hobe, M. *et al.* (2001) A molecular link between stem cell regulation and floral patterning in *Arabidopsis*. *Cell* **105**, 793–803.
Parcy, F., Nilsson, O., Busch, M.A., Lee, I. & Weigel, D. (1998) A genetic framework for floral patterning. *Nature* **395**, 561–566.
Wagner, D., Sablowski, R.W. & Meyerowitz, E.M. (1999) Transcriptional activation of *APETALA1* by *LEAFY*. *Science* **285**, 582–584.

LEUNIG

Conner, J. & Liu, Z. (2000) *LEUNIG*, a putative transcriptional corepressor that regulates *AGAMOUS* expression during flower development. *Proceedings of the National Academy of Science, USA* **97**, 12902–12907.

Case study 5.5: Adaxial–abaxial axis of the *Antirrhinum* flower

For a general review see:
Coen, E.S. (1996) Floral symmetry. *EMBO Journal* **15**, 6777–6788.

CYCLOIDEA and DICHOTOMA

Luo, D., Carpenter, R., Copsey, L. *et al.* (1999) Control of organ asymmetry in flowers of *Antirrhinum. Cell* **99**, 367–376.

Luo, D., Carpenter, R., Vincent, C., Copsey, L. & Coen, E. (1996) Origin of floral asymmetry in *Antirrhinum. Nature* **383**, 794–799.

DIVARICATA

Almeida, J., Rocheta, M. & Galego, L. (1997) Genetic control of flower shape in *Antirrhinum majus. Development* **124**, 1387–1392.

CENTRORADIALIS

Bradley, D., Carpenter, R., Copsey, L. *et al.* (1996) Control of inflorescence architecture in *Antirrhinum. Nature* **379**, 791–797.

CHAPTER 6

Position relative to a particular cell, tissue or organ

Chapters 4 and 5 discussed how cell fate is regulated by axial information in the plant. Theoretically, axial information alone could ensure that every cell adopted a fate appropriate to its position. However, this is not the case and plants use additional mechanisms to coordinate cell fate with cell position. This chapter discusses the important example of the regulation of cell fate according to the proximity of a particular cell, tissue or organ. Proximity-based information ensures that certain structures, such as individual leaves, always develop apart from each other; while other structures, such as a leaf and its vascular supply, are always connected, producing a functional plant.

There are theoretical similarities between proximity-based mechanisms acting in plant development and those that function in animal embryos. In animal embryology, a distinction is made between **equivalent** and **non-equivalent cells**. In a group of equivalent cells (an **equivalence group**) all cells have equal developmental potential. In contrast, non-equivalent cells have different developmental potentials. Cellular interactions during animal development are categorized according to whether they occur within an equivalence group or between non-equivalent cells. Interactions within an equivalence group come under the broad heading of **lateral specification**. An interaction in which the fate of one cell depends upon a signal from a non-equivalent cell is called **induction**.

It is unclear whether the terms 'equivalent' and 'non-equivalent' are meaningful when considering plant cells. As discussed in Chapter 3, plant cells are not restricted to particular fates by their lineage and, therefore, cells in a developing organ share the same developmental potential. However, the concepts of lateral specification and induction are useful when discussing cellular interactions during plant development.

Lateral specification allows patterns to arise from initially uniform groups of cells. For example, case study 6.1 considers how lateral specification may generate the array of trichomes on the *Arabidopsis* leaf from an initially uniform field of epidermal cells. Induction allows the elaboration of existing patterns. For example, in the *Arabidopsis* root, the position of an epidermal cell relative to cells in the underlying root cortex determines whether the epidermal cell will develop a root hair. This suggests that root hair development in *Arabidopsis* is controlled by an inductive signal from deeper layers of the root. The pattern of root hairs is discussed in case study 6.2.

Proximity-based mechanisms can also generate patterns on a larger scale. For example, there is good evidence that the arrangement of leaves on the shoot—the **phyllotaxy**—is controlled through the influence of existing leaf primordia on subsequent leaf initiation. This is discussed in case study 6.3. Lastly, case study 6.4 discusses vascular development: in particular, whether auxin exported by leaf primordia induces the differentiation of vascular connections between leaves and the stem.

Case study 6.1: The pattern of trichomes on the *Arabidopsis* leaf

Arabidopsis trichomes (leaf hairs) are single epidermal cells that project out from the epidermis and usually have three branches (Fig. 6.1). Viewed from above, the branches form a 'Y' shape pointing towards the leaf base, suggesting that the branching pattern responds to axial information in the leaf primordium. Trichome distribution varies between leaves on the shoot and between the upper and lower epidermis of individual leaves. In each region of epidermis that contains trichomes, however, the trichomes are more evenly spaced than chance would dictate. Statistical analysis shows that trichomes develop in pairs and clusters far less often than expected for a random distribution. An arrangement such as this, where there is a minimum distance between individual structures, is called a **spacing pattern**. This case study considers how the trichome spacing pattern arises.

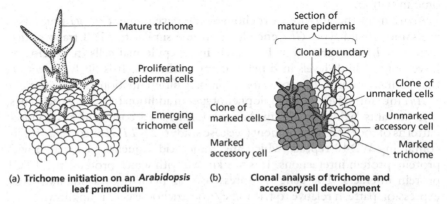

(a) **Trichome initiation on an *Arabidopsis* leaf primordium**

(b) **Clonal analysis of trichome and accessory cell development**

Fig. 6.1 Trichome development in *Arabidopsis*. (a) Trichome initiation on a leaf primordium. The first round of trichome initiation occurs in a basipetal (tip-to-base) sequence. (b) Clonal analysis of trichome and accessory cell development. The central trichome belongs to the marked clone (pink) but some of its accessory cells are unmarked (white). Therefore, the trichome and the unmarked accessory cells belong to different lineages. (Adapted in part from Schnittger *et al.*, 1999.)

Clonal analysis

Trichomes begin to differentiate early in leaf development while cells in the surrounding epidermis are still dividing (Fig. 6.1). As the leaf primordium grows, the existing trichomes become more widely separated and further rounds of trichome initiation occur in the gaps between them. In the later stages of leaf development, epidermal cell divisions and leaf growth occur without trichome initiation and the distances between neighbouring trichomes are increased. Finally, the epidermal cells immediately surrounding each trichome enlarge to become **accessory cells**.

The very rare occurrence of trichome pairs could be explained if each trichome and its surrounding accessory cells all descend from a single mother cell: then, although trichome mother cells might be neighbours, trichomes would always be separated. However, clonal analysis of the *Arabidopsis* leaf indicates that a trichome and its accessory cells are not always clonally related (Fig. 6.1b). Consequently, a clonal mechanism is unlikely to control trichome distribution. The obvious alternative is that the trichome pattern results from signalling between epidermal cells.

Genetics of trichome patterning

A simple model to explain how cell-to-cell signalling could generate the trichome spacing pattern has been proposed on the basis of mutational analysis. Mutations in at least 20 genes disrupt *Arabidopsis* trichome development. Most of these affect aspects of trichome differentiation but loss-of-function mutations in either *GLABRA1*—sometimes called *GLABROUS1*—(*GL1*) or *TRANSPARENT TESTA GLABRA1* (*TTG1*) result in almost entirely hairless leaves (Fig. 6.2). Therefore, *GL1* and *TTG1* are both required for trichome initiation.

Apart from the absence of trichomes, the development of *gl1* mutants is the same as wild type. *GL1* encodes a protein similar to MYB transcription factors. *GL1* is expressed at low levels in all epidermal cells in developing leaves and at high levels in differentiating trichomes. It is likely, therefore, that an increase in *GL1* expression promotes differentiation as a trichome.

ttg1 mutants have a pleiotropic phenotype. In addition to lacking trichomes, *ttg1* mutants lack anthocyanin, have abnormal seed coats and have altered patterns of root hair development (see case study 6.2). *TTG1* encodes a protein containing WD40 repeats, which are amino acid sequences associated with protein–protein interactions. It is not known with which proteins the TTG1 protein interacts. The gene is expressed in all plant organs but its precise expression pattern relative to trichome/non-trichome cells is unknown.

Interestingly, the *ttg1* mutation can be complemented by transgenic expression of the maize *R* gene (Fig. 6.2d) (like *TTG1*, *R* functions in controlling anthocyanin synthesis). *ttg1* mutants that contain a constitutively expressed *R* transgene produce abundant trichomes. This suggests that an *Arabidopsis R* homologue acts at the same point as, or downstream of, *TTG1*

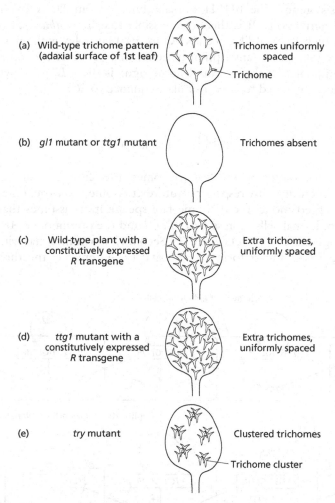

Fig. 6.2 The effects of mutations and transgene expression on trichome pattern in *Arabidopsis*. The trichomes are viewed from above so that their branches appear as a 'Y' shape, and their stalks cannot be seen. (a) On wild-type leaves, a minimum distance is usually maintained between neighbouring trichomes. (b) Trichomes do not develop on the leaves of *gl1* and *ttg1* mutants, indicating that *GL1* and *TTG1* are required for trichome fate. (c) Constitutive expression of an *R* transgene in a wild-type background results in the production of abundant, uniformly spaced trichomes. (d) Constitutive expression of an *R* transgene in a *ttg1* mutant also results in abundant, uniformly spaced trichomes, indicating that *R* expression overcomes the block to trichome formation resulting from loss of *TTG1* function. (e) Trichomes frequently develop in clusters on the leaves of *try* mutants, indicating that *TRY* is required for trichome spacing.

in controlling trichome initiation. *R* expression cannot overcome the block to trichome initiation imposed by the *gl1* mutation. Likewise, transgenic plants that constitutively express *GL1* are still bare of trichomes if they are mutant for *ttg1*. Therefore, trichome initiation requires a combination of *GL1* and *TTG1* or *R* expression.

R encodes a basic helix–loop–helix (bHLH) protein that interacts with MYB transcription factors in maize to regulate the transcription of anthocyanin biosynthesis genes (the bHLH domain functions in DNA binding and protein dimerization). It is therefore possible that in *Arabidopsis*, the GL1 protein (which is MYB-like, see above) forms a complex with an *R* homologue to regulate the transcription of genes that promote trichome formation. A candidate for the role of *R* homologue is the *GL3* gene, which has recently been reported to have a similar sequence to *R*.

A model for trichome spacing

Prior to the initiation of the first trichomes, the leaf epidermis acts as an equivalence group with respect to future trichome/non-trichome fate. A widely accepted model for trichome fate specification assumes that at this stage all epidermal cells express *GL1*, *TTG1* and the presumptive *Arabidopsis* *R* homologue at a low level (Fig. 6.3). Secondly, it assumes that within each cell, *GL1*, *TTG1* and *R* form an autocatalytic system, i.e. the three genes

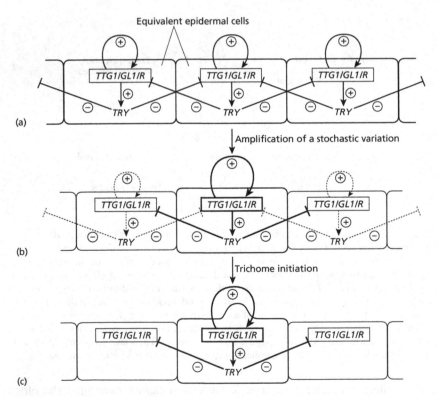

Fig. 6.3 A model for trichome spacing in *Arabidopsis*. (a) Initially a state of mutual inhibition exists between leaf epidermal cells with respect to trichome development. (b, c) Stochastic variations in gene expression are amplified to commit individual cells to trichome development and to prevent neighbouring cells from following suit. Arrows accompanied by a 'plus' sign indicate positive regulation. Lines ending in a bar and accompanied by a 'minus' sign indicate negative regulation. (Adapted from Schnittger *et al.*, 1999.)

act in concert to increase their own expression. The third assumption of the model is that when a threshold of *GL1–TTG1–R* expression is reached, the cell differentiates as a trichome. This would give all epidermal cells a tendency towards trichome fate.

To achieve the spacing pattern, there must also be an inhibitor of trichome fate. This function appears to be provided by the *TRIPTYCHON (TRY)* gene. *try* mutants develop an approximately wild-type number of trichomes but a significant proportion of the trichomes occur in clusters (Fig. 6.2e). This suggests that *TRY* is required to inhibit trichome clustering. The model proposes that *GL1–TTG1–R* expression promotes individual cells to express *TRY*, and that *TRY* acts to reduce *GL1–TTG1–R* expression in neighbouring cells.

Trichome spacing is therefore determined by two feedback loops: a cell-autonomous, positive feedback loop ratchets up *GL1–TTG1–R* expression; whilst a negative feedback loop downregulates *GL1–TTG1–R* in neighbouring cells (Fig. 6.3). According to the model, in the very young leaf primordium, all epidermal cells express *GL1–TTG1–R*, and therefore *TRY*, at relatively low levels. Consequently, neighbouring cells are caught in a pattern of mutual inhibition. *TRY* expression and hence inhibitor production by all members of a local group of cells prevents *GL1–TTG1–R* expression in any one cell from passing the threshold for trichome differentiation. However, this equilibrium is unstable and is disrupted by stochastic variations in gene expression between cells.

If, for example, a cell acquires a slightly higher level of *GL1–TTG1–R* expression than its neighbours, then that cell will subsequently have slightly higher *TRY* activity and will produce more inhibitor. The increase in inhibitor production will reduce *GL1–TTG1–R* expression in neighbouring cells leading to a reduction in *TRY* expression and inhibitor production by those cells. Exposed to less of the inhibitor, the cell with the initially higher *GL1–TTG1–R* activity will increase *GL1–TTG1–R* expression still further. Therefore, the initial stochastic variation in gene expression is amplified until one cell is selected to become a trichome, and its immediate neighbours are prevented from following suit.

A stochastic selection process is only required during the initiation of the first trichomes on the leaf. As the leaf primordium grows and the first trichomes spread out, cells at a sufficient distance from existing trichomes will receive less of the proposed inhibitory signal and will therefore be free to develop into new trichomes. Each new trichome, of course, will inhibit its immediate neighbours from also differentiating as trichomes. Such a process, where like inhibits like, is known as **lateral inhibition**. Developing trichomes are separated by only three to four epidermal cells, therefore the inhibitor need act over only a few cell diameters.

Spacing mechanisms in which a locally acting, autocatalytic promoter of cell fate induces the production of a longer range inhibitor of that fate, were first proposed in the early 1950s by the mathematician Alan Turing. **Turing mechanisms** (also called **reaction–diffusion mechanisms** since they can be modelled as promotion/inhibition interactions between diffusible substances) have been implicated in a number of developmental systems, for example the spacing of sensory bristles on *Drosophila* (the fruit fly).

Case study 6.2: The pattern of root hairs in *Arabidopsis*

Cells in the *Arabidopsis* root epidermis differentiate either as root hair cells (**trichoblasts**) or as hairless cells (**atrichoblasts**). This cell fate decision depends on the position of each epidermal cell relative to the underlying root cortex (see below). Therefore, epidermal cell fate is likely to be regulated by an inductive signal originating in deeper layers of the root.

Trichoblast/atrichoblast development

The *Arabidopsis* root epidermis is derived from a ring of 16 epidermal/lateral root cap initials in the root apical meristem (Fig. 1.5 shows the structure of the *Arabidopsis* root tip). Behind these initials, transverse divisions in the epidermis produce regular files of cells running along the root, and occasional longitudinal divisions increase the number of files from 16 to about 20. In the cortex, which in the *Arabidopsis* root is a single cell layer immediately beneath the epidermis, there are only eight cell files. Epidermal cells that lie over the cleft between two cortical cells develop as trichoblasts, those that lie over the outer wall of a single cortical cell develop as atrichoblasts. Since epidermal and cortical cell files lie approximately parallel to each other along the root, the epidermis consists of eight files of trichoblasts which are separated from each other by one or two files of atrichoblasts (Fig. 6.4).

The development of trichoblasts differs from that of atrichoblasts from a very early stage. As discussed in Chapter 1, the root tip is roughly divided into a zone of cell division, a zone of cell expansion and a maturation zone: each zone lying progressively farther from the apical meristem. Immediately behind the apical meristem, in the division zone, presumptive trichoblasts have a denser cytoplasm and are less vacuolated than presumptive atrichoblasts (Fig. 6.4). During subsequent development, cells in trichoblast files divide more often and elongate less than cells in atrichoblast files. Consequently, as cells enter the maturation zone, trichoblast files consist of more numerous but shorter cells than atrichoblast files. In the maturation zone, trichoblasts but not atrichoblasts produce root hairs.

The position of cells in the root epidermis can be altered by laser ablation. As discussed in case study 3.1, ablated cells collapse and neighbouring cells are displaced into the resulting gap. When a presumptive hair cell moves into the gap left by an ablated non-hair cell, the relocated cell switches to a non-hair cell fate. In the reverse case, a cell from an atrichoblast file will switch to a trichoblast fate if forced into a trichoblast position. Cell fate in the root epidermis is therefore determined by position rather than lineage, consistent with the evidence presented in Chapter 3.

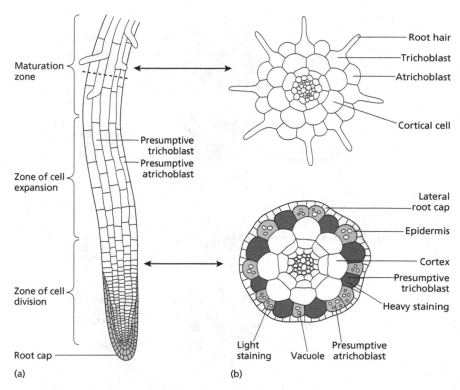

Fig. 6.4 The development of trichoblasts and atrichoblasts in *Arabidopsis* roots. (a) The *Arabidopsis* root tip showing the zones of cell division, expansion and maturation (a section of root cap has been omitted to reveal the underlying epidermis). Note that in the zone of cell expansion, trichoblasts elongate less than atrichoblasts. (b) Radial cross-sections through the root showing the arrangement of trichoblasts and atrichoblasts in the maturation zone (top), and the features of presumptive trichoblasts and atrichoblasts in the zone of cell division (bottom). ((b) adapted from Dolan *et al.*, 1993; Lee & Schiefelbein, 1999.)

Genetics of the trichoblast/atrichoblast decision

Mutational analysis of root hair development has identified genes that promote atrichoblast fate over trichoblast fate, and also genes that promote trichoblast fate over atrichoblast fate. This section discusses four genes that promote atrichoblast development, and one gene that promotes trichoblast development. It then outlines a model for the determination of epidermal cell fate.

TRANSPARENT TESTA GLABRA1, R and WEREWOLF promote atrichoblast development

As discussed in case study 6.1, *TTG1* encodes a protein containing WD40 repeats (amino acid sequences associated with protein–protein interactions) and is required for trichome initiation in the shoot. In *ttg1* mutants almost every root epidermal cell develops as a trichoblast (Fig. 6.5b). In the

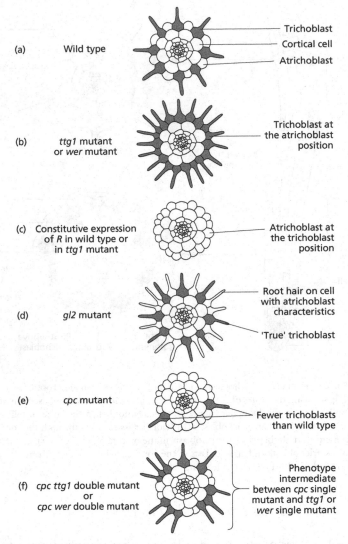

Fig. 6.5 The effects of mutations and transgene expression on the development of the *Arabidopsis* root epidermis. The figure shows radial cross-sections through the root in the maturation zone. Cells displaying full trichoblast characteristics are coloured red. (a) In the wild-type root, epidermal cells lying over the anticlinal walls separating two cortical cells differentiate as trichoblasts, while cells lying over the periclinal wall of a single cortical cell differentiate as atrichoblasts. (b) In *ttg1* and *wer* mutants, trichoblasts develop in place of atrichoblasts, indicating that *TTG1* and *WER* are required for atrichoblast fate. (c) Constitutive expression of an *R* transgene results in an almost hairless root in both a wild-type and a *ttg1* mutant background. This indicates that *R* promotes atrichoblast fate and that *R* expression can compensate for the loss of *TTG1* function in the specification of atrichoblasts. (d) In *gl2* mutants, cells in the atrichoblast position produce root hairs but otherwise have atrichoblast characteristics, indicating that *GL2* is required specifically to prevent root hair outgrowth. (e) In *cpc* mutants, the number of trichoblasts is reduced, indicating that *CPC* acts to promote trichoblast fate. (f) *cpc ttg1* and *cpc wer* double mutants have phenotypes intermediate between *cpc* single mutants and *ttg1/wer* single mutants. This is consistent with an antagonistic relationship in which *CPC* promotes trichoblast fate, and *TTG1* and *WER* promote atrichoblast fate.

root, therefore, *TTG1* is required for the development of atrichoblasts—paradoxically, the gene is needed for 'hair' cells in the leaf, but hairless cells in the root.

Also as discussed in the last case study, many of the phenotypic effects of the *ttg1* mutation can be counteracted by constitutive expression of the maize *R* gene. Consistent with this, the root epidermis of *ttg1* mutants in which *R* is constitutively expressed is almost devoid of root hairs (Fig. 6.5c), the opposite of the *ttg1* phenotype. This suggests that an *Arabidopsis R* homologue—possibly *GL3* (see case study 6.1)—acts either at the same point as, or downstream of *TTG1*. Constitutive *R* expression also blocks trichoblast development on wild-type roots. So, like *TTG1*, *R* promotes atrichoblast over trichoblast development. As stated in case study 6.1, *R* encodes a bHLH protein. Such proteins can act in concert with MYB transcription factors to regulate gene expression.

A third gene that promotes atrichoblast development is *WEREWOLF* (*WER*). As the name suggests, *wer* mutants have very hairy roots (Fig. 6.5b). The *wer* phenotype is very similar to the *ttg1* phenotype: trichoblasts develop in place of atrichoblasts. *WER* encodes a MYB-like transcription factor which has been shown (in a yeast system) to interact directly with the bHLH protein encoded by *R*. This suggests that the WER protein may interact with an *Arabidopsis R* homologue to regulate gene transcription in root epidermal cells. *WER* expression is higher in atrichoblast files than in trichoblast files, consistent with a role in promoting atrichoblast fate.

GLABRA2 prevents production of hairs on atrichoblasts

Ectopic root hair cells on *ttg1* and *wer* mutants have all the characteristics of trichoblasts. In contrast, on *glabra2* (*gl2*) mutants, ectopic root hairs grow from cells that otherwise resemble atrichoblasts: with the atrichoblast pattern of vacuolation, elongation and division (Fig. 6.5d). The development of true trichoblasts on *gl2* roots is normal. Therefore, *GL2* is only needed at the final stage of atrichoblast development, at which time it acts to prevent the formation of a root hair.

GL2 encodes a homeodomain transcription factor and is preferentially expressed in atrichoblast files. Presumably, the GL2 protein regulates gene transcription to prevent the formation of a root hair. Although *GL2* does not appear to function until the final stages of atrichoblast development, its preferential expression in atrichoblast files is detectable even in the division zone of the root.

Position-dependent expression of *GL2* is abolished in *wer* mutants, in which *GL2* expression is greatly reduced and occurs in randomly scattered cells. It is likely, therefore, that one function of *WER* is to promote *GL2* expression in atrichoblasts. *GL2* expression is also affected by *TTG1* and *R*. Expression is greatly reduced in *ttg1* mutants but in this case the spatial pattern of *GL2* activity is unaffected, *GL2* is still more active in the atrichoblast position. As expected by the hairless phenotype of roots with constitutive *R* expression (see above) such roots express *GL2* in all epidermal cells.

In summary, *TTG1*, *R* and *WER* are required throughout atrichoblast development, and *WER* at least is preferentially expressed in developing atrichoblasts. One target of these three genes is *GL2* which is required late in atrichoblast development to prevent the formation of a root hair.

CAPRICE promotes root hair development

The loss-of-function *caprice* (*cpc*) mutation results in roots that have about four times fewer root hairs than wild type (Fig. 6.5e), whereas constitutive expression of a *CPC* transgene leads to root hairs developing from almost all cells in the root epidermis. Therefore, *CPC* expression promotes root hair development.

The functional relationship between *CPC* and the genes that promote atrichoblast development (above) has been investigated by generating double mutants (Fig. 6.5f). *cpc ttg1* and *cpc wer* double mutants have root hair densities intermediate between those of the respective single mutants. This is consistent with an antagonistic relationship involving the promotion of atrichoblast fate by *TTG1* and *WER*, and the promotion of trichoblast fate by *CPC*. In contrast to these intermediate phenotypes, *cpc gl2* double mutants have root hairs on almost every epidermal cell, i.e. they resemble *gl2* mutants. Therefore, the loss of root hairs in *cpc* mutants requires *GL2* function.

A model for the control of root hair development

CPC encodes a protein containing a DNA-binding domain homologous to those in MYB-type transcription factors. However, the CPC protein lacks a MYB-type transcriptional activator domain. The nature of the CPC protein, together with the double mutant phenotypes described above, has led to a simple model for the control of root hair outgrowth (Fig. 6.6).

The model assumes that the WER protein and the CPC protein, both MYB-like, compete to interact with the bHLH protein encoded by an *Arabidopsis R* homologue. A WER–R protein complex forms a functional transcriptional activator that induces *GL2* transcription and so prevents root hair formation. Consistent with this idea, the *GL2* promoter contains a putative MYB-binding site. A CPC–R complex is inactive because CPC lacks a transcriptional activator domain and therefore *GL2* is not transcribed and a root hair develops. The pattern of root hairs is determined by the balance of WER and CPC activity, with WER dominating in atrichoblast files, and CPC dominating in trichoblast files.

This model explains why constitutive overexpression of *R* results in a hairless root. When R protein is abundant, CPC levels are insufficient to prevent WER–R complex formation in any cell file. The precise role of *TTG1* is unclear. As discussed above, *TTG1* regulates *GL2* expression since loss of *TTG1* function reduces the level of *GL2* transcription. Also, *TTG1* must act upstream of, or at the same point as, *R* since constitutive *R* expression reverses the *ttg1* phenotype.

A potential problem with the model is that *CPC* appears to be expressed primarily in atrichoblasts. Given that CPC promotes trichoblast development,

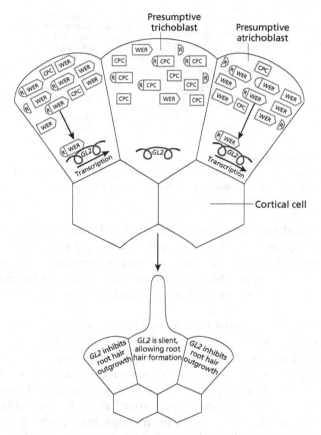

Fig. 6.6 A model for the control of root hair outgrowth. In this model, WER and CPC, which are both MYB-like proteins, compete to bind to the bHLH protein encoded by an *Arabidopsis R* homologue. In presumptive atrichoblasts, WER is more common than CPC, allowing the formation of WER–R dimers. These induce transcription from the *GL2* gene, which inhibits root hair outgrowth. In presumptive trichoblasts, CPC is more common than WER, preventing the formation of WER–R dimers. Since CPC lacks a transcriptional activator domain, the CPC–R dimer does not induce *GL2* transcription, and consequently the cell is able to form a root hair. (Adapted from Lee & Schiefelbein, 1999.)

this is the opposite cell type to that expected. If this expression pattern is confirmed, then *CPC* acts as a non-cell-autonomous regulator of trichoblast fate. It is possible that the CPC protein is transported from atrichoblasts to trichoblasts via plasmodesmata. With this addition, the model described above resembles lateral inhibition. CPC exported from atrichoblasts would prevent cells in trichoblast files from adopting atrichoblast fate.

Effects of ethylene on root hair development

The relationship between root hair development and the position of epidermal cells relative to the root cortex suggests the involvement of signals dependent on the root architecture. Ethylene is probably one such signal.

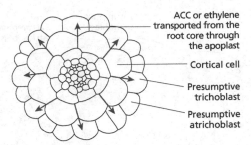

Fig. 6.7 The possible influence of *Arabidopsis* root architecture on ethylene movement. The anticlinal walls of cells in the cortex provide a direct apoplastic path from the root core to presumptive trichoblasts. Transport of ethylene or its precursor ACC along this path (red arrows) may trigger root hair outgrowth.

Roots treated with ethylene or with its biochemical precursor ACC (1–amino-1-cyclopropane carboxylate) are very hairy, having hairs on cells in both the trichoblast and atrichoblast position. Similarly, *constitutive triple response1* (*ctr1*) mutants, which have a constitutively active ethylene response, produce excessive root hairs. Therefore, ethylene promotes root hair development.

As in *gl2* mutants, perturbations to ethylene signalling affect root hair outgrowth without changing earlier differences in the behaviour of cells in the trichoblast and atrichoblast positions. Furthermore, when presumptive trichoblasts and atrichoblasts enter the maturation zone, they are not equally responsive to ethylene. Although many ectopic root hairs develop on *ctr1* plants, there are still substantially more root hairs in the trichoblast cell files than in the atrichoblast files. Cells in the trichoblast position are therefore more competent than cells in the atrichoblast position to produce a root hair in response to ethylene signalling. Earlier controls on trichoblast/atrichoblast development may 'prime' trichoblasts to respond to ethylene by producing a root hair, or inhibit this response in atrichoblasts, or both.

The effects of ethylene may contribute to root hair patterning in *Arabidopsis* (Fig. 6.7). Epidermal cells lying over the cleft between two cortical cells are connected by a direct apoplastic path to deeper layers of the root. ACC or ethylene synthesized in the core of the root might preferentially reach these cells and induce root hair formation.

Relationship between root hair and trichome development

TTG1, *R* and *GL2* specify trichome development in the shoot (see case study 6.1) and atrichoblast development in the root. In both atrichoblast and trichome development, *GL2* acts downstream of *TTG1* and *R*. *TTG1*, *R* and *GL2* may therefore represent a developmental module that is re-used in different aspects of epidermal development. The overlap between the mechanisms acting in atrichoblast development and in trichome development is not absolute. The *gl1* and *try* mutations affect trichome development but mutants have normal root hair development. Conversely, the *wer* and *cpc* mutations affect root hair but not trichome development. As discussed in this case study and the last, GL1, WER and CPC are all MYB-like proteins, and may all represent partners for bHLH proteins encoded by one or more *Arabidopsis R* homologues.

The pattern of stomata on the hypocotyl

In the hypocotyl, the relationship between epidermal cell files and cortical cell files is essentially the same as that in the root. The hypocotyl does not produce root hairs but does form stomata. These develop preferentially in epidermal files overlying the intersection between two cortical cell files, the equivalent of the trichoblast position in the root.

Interestingly, mutations that alter root hair patterning cause equivalent alterations in the patterning of stomata on the hypocotyl. For example, *ttg1* mutants develop trichoblasts in atrichoblast positions on the root, and also stomata in the equivalent, non-stomatal positions on the hypocotyl. This suggests that the mechanism that generates the pattern of root hairs is re-used to generate the pattern of stomata in the hypocotyl. The pattern of stomata in the rest of the plant is controlled by a different mechanism and remains normal in mutants with altered root hair patterning.

Case study 6.3: Phyllotaxy

The previous two case studies discussed patterns of cell types. This case study considers patterning on a larger scale, taking as an example the mechanisms that control the arrangement of leaves on the shoot, i.e. the **phyllotaxy**. The case study begins by describing a range of common phyllotaxies. It then discusses the possible mechanisms controlling phyllotactic patterns, as revealed by surgical experiments and inductive treatments.

Patterns of leaf initiation

As described in Chapter 1, the shoot apical meristem is divided radially into a **central zone** of slowly dividing cells and a **peripheral zone** of more rapidly dividing cells. Leaf primordia are initiated in the peripheral zone to produce a succession of leaf-bearing **nodes** separated by leafless **internodes**.

In most plants a single leaf develops at each node and successive leaves form a spiral running up the shoot. This is **spiral phyllotaxy** (Fig. 6.8a). In shoots with **alternate phyllotaxy**, a single leaf develops at each node and successive leaves form on opposite sides of the stem (Fig. 6.8b). In some plants, two leaves arise at each node, normally opposite one another. If each leaf pair forms directly above the preceding pair, the phyllotaxy is **distichous**. If each leaf pair develops at right angles to the preceding pair, the phyllotaxy is **decussate** (Fig. 6.8c). In shoots with **whorled phyllotaxy**,

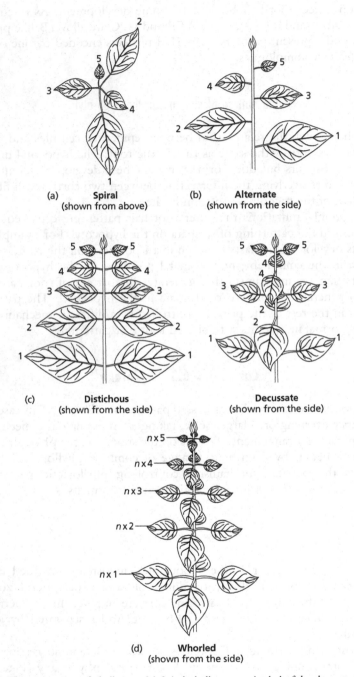

Fig. 6.8 The main types of phyllotaxy. (a) Spiral phyllotaxy: a single leaf develops at each node and successive leaves are arranged in a spiral pattern. This is the most common type of phyllotaxy. (b) Alternate phyllotaxy: a single leaf develops at each node and successive leaves are produced on alternate sides of the shoot. (c) Distichous and decussate phyllotaxy: a pair of leaves is produced at each node. In plants with distichous phyllotaxy, each leaf pair forms directly above the previous pair. In plants with decussate phyllotaxy, each leaf pair is oriented perpendicular to the previous pair. (d) Whorled phyllotaxy: a whorl of leaves develops at each node.

more than two leaves form at each node (Fig. 6.8d). Depending on the species, successive whorls may have the same angular position or may be rotated relative to each other.

Surgical experiments on phyllotaxy

In all phyllotaxies, the positions of new leaf primordia can be predicted from the sites of existing primordia. By implication, existing leaf primordia influence the sites of leaf initiation. This hypothesis has been well supported by classic surgical experiments performed on the shoot apical meristem.

Lupin, for example, has spiral phyllotaxy such that each leaf primordium is flanked by two older primordia. As described in Chapter 5, existing leaf primordia are described as P1, P2, P3, etc. in order of ascending age. Future primordia are described as I1, I2, I3, etc., with I1 being the next primordium to appear. In lupin, I1 develops between P2 and P3, whereas I2 appears between P1 and P2 (Fig. 6.9a). The influence of existing primordia on phyllotaxy was investigated by making an incision that isolated P1 from more central regions of the shoot apex. Such an incision does not affect the

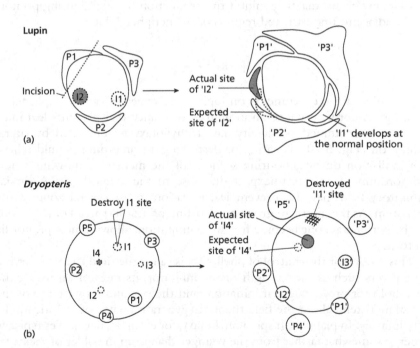

Fig. 6.9 The effects of isolating or destroying a leaf primordium on the positioning of subsequent primordia. Primordia whose positions are affected by the experiments are coloured red. (a) Isolating P1 from the centre of the lupin shoot apical meristem results in a change in the position of I2. On treated apices, I2 develops closer to P1 than normal. (b) Destroying the I1 site on the *Dryopteris* shoot apical meristem results in a change in the position of I4. On treated apices, I4 develops closer to the I1 site than normal. These results indicate that existing leaf primordia inhibit the initiation of new leaf primordia in immediately neighbouring regions of the shoot apical meristem. (Adapted from Wardlaw, 1949; Steeves & Sussex, 1989.)

location of I1 (which is not adjacent to P1) but does alter the position of I2. I2 now forms closer to P1 and farther from P2 than expected (Fig. 6.9a).

The experiments on lupin were originally interpreted in terms of the space available for leaf initiation. Lupin leaf primordia are large relative to the apical meristem and it was suggested that new primordia emerge in the first available gap between existing primordia. In the experiment described above, the incision restricts the growth of P1 hence there is more room in the adjacent section of the meristem for I2 to appear.

Subsequent experiments, however, led to a different interpretation. Leaf primordia in the fern *Dryopteris dilatata* are small compared to the apical meristem and so a phyllotactic mechanism based on restricted space is unlikely. Despite this, surgical experiments on *Dryopteris* have similar results to those on lupin. Like lupin, *Dryopteris* has spiral phyllotaxy, although the pattern of primordia is different. In *Dryopteris*, I1 grows between P2 and P3, I2 forms between P1 and P2, I3 develops between P1 and P3, and I4 appears between I1 and P2 (Fig. 6.9b). If the I1 site is surgically destroyed, the positions of I2 and I3 are unaffected. In contrast, I4, which is normally flanked on one side by I1, now forms unusually close to the I1 site (Fig. 6.9b). This indicates that existing primordia do more than simply get in the way of new primordia; they inhibit the production of neighbouring primordia in adjacent, unobstructed regions of the peripheral zone.

The field theory

The results of the experiments on lupin and *Dryopteris* can be explained if existing primordia produce an inhibitory substance that prevents leaf initiation in their immediate vicinity, i.e. if phyllotaxy is controlled by lateral inhibition (Fig. 6.10). Isolating or destroying a primordium would relieve the inhibition on neighbouring sections of the meristem, allowing a new primordium to emerge unexpectedly close to the surgical site. A similar inhibitory field may also prevent leaf initiation in the central zone of the meristem. As the meristem grows and existing primordia spread out, sections of the peripheral zone become free from inhibition, allowing new primordia to form.

This model for the control of phyllotaxy is called the **field theory**. Varying parameters such as the strength of the inhibitor, its rates of diffusion, the threshold for preventing leaf initiation, and the size and growth rate of the apical meristem, allows the field theory to generate a range of realistic model phyllotaxies. In plants with spiral phyllotaxy, for example, new leaves typically emerge somewhat farther from the younger than from the older of their two flanking primordia. This can be explained by assuming that the strength of inhibition around a primordium decreases as the primordium ages.

The field theory is generally accepted but it remains unproven. There is now some suggestive genetic data. The *terminal ear1* (*te1*) mutation of maize results, among several other things, in frequent deviations from the wild-type alternate phyllotaxy. Interestingly, *TE1* is expressed in successive horseshoe-shaped regions at the shoot apex, with the open ends of the

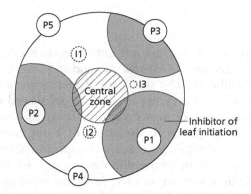

Fig. 6.10 The field theory of phyllotaxy applied to the *Dryopteris* apex. The field theory proposes that the central zone of the shoot apical meristem and existing leaf primordia produce inhibitors of leaf initiation, resulting in inhibitory fields at the centre of the meristem and around each leaf primordium. New leaf primordia are initiated in the gaps that arise between these fields. The figure shows the hypothetical inhibitory fields (pink) surrounding the central zone and the youngest three leaf primordia on a *Dryopteris* shoot apical meristem, and the sites at which I1, I2 and I3 will develop.

'horseshoes' aligned with the midrib-forming region of successive leaf primordia, and the uppermost 'horseshoe' aligned with I1 (Fig. 6.11). The midrib-forming region of the maize leaf primordium is the first section of the primordium to emerge from the shoot apical meristem, suggesting that leaf primordia are initiated where *TE1* expression is lowest. *TE1* expression may therefore mark the zone in which an inhibitor of leaf initiation is active. *TE1* encodes a putative RNA-binding protein but its precise biochemical function is unknown.

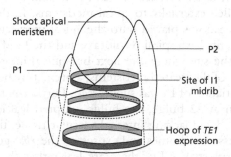

Fig. 6.11 *TE1* expression in the maize shoot apex. The *TE1* gene of maize is expressed in successive horseshoe-shaped regions at the shoot apex (red). The open end of each 'horseshoe' is aligned with the midrib-forming region of the adjacent leaf primordium, and the uppermost 'horseshoe' is associated with I1. Since the midrib-forming region will be the first section of the I1 primordium to emerge from the apical meristem, *TE1* expression may mark the zone in which an inhibitor of leaf initiation is active. (Adapted from Veit *et al.*, 1998.)

Auxin

Blocking auxin transport prevents leaf initiation. This can be seen both in transport-deficient mutants of *Arabidopsis* and in plants treated with polar auxin transport inhibitors. The shoot apical meristems of these plants are active, but fail to initiate leaves and so produce pin-shaped shoots.

A series of experiments on tomato apices shows that auxin induces leaf formation. If auxin is applied to the flanks of a pin-shaped tomato apex generated by treatment with a polar auxin transport inhibitor, a leaf primordium forms. The size of the primordium is proportional to the amount of auxin applied. Auxin also promotes leaf initiation when applied to normal tomato meristems. Auxin applied to the I1 site causes an enlarged primordium to form, and auxin applied to the I2 site causes premature leaf initiation. Exogenous auxin cannot induce leaf initiation from the centre of the meristem or from below a certain point on its flanks. This indicates that only cells in the peripheral zone are competent to initiate leaves in response to auxin.

These results suggest a relationship between leaf initiation and auxin, but their significance for understanding the control of phyllotaxy is unclear. The field theory proposes that leaf primordia produce an inhibitor of leaf initiation. However, leaf primordia are thought to be major sites of auxin synthesis, which is a leaf promoter. This paradox remains unresolved.

Expansins

Expansins are extracellular proteins that increase the extensibility of the cell wall. They are encoded by a family of genes whose members display a variety of expression patterns. In the tomato shoot apical meristem, the expression of at least one expansin gene rises at the I1 site. This suggests that the initiation of a leaf primordium is associated with an increase in wall extensibility. To investigate the significance of wall extensibility during leaf initiation, expansin was applied externally to apical meristems. In this experiment, an expansin-loaded bead was placed onto the apical meristem of tomato (Fig. 6.12). Tomato plants have spiral phyllotaxy and the bead was put onto the I2 position, i.e. the site where the next-but-one primordium was due to emerge. On some of the treated meristems, a premature bulge formed at the I2 site and the initiation of I1 was suppressed. Furthermore, in a proportion of the plants with an I2 bulge, the bulge adopted leaf identity. Although roughly peg-shaped and lacking internal vascular tissue, the leaf-like bulges were green, possessed trichomes and expressed the *rbcS* gene (encoding the small subunit of rubisco). All of these are leaf rather than apical meristem characteristics. In meristems where the I2 bulge displayed leaf-like features, the bulge influenced subsequent leaf initiation, resulting in the reversal of the phyllotactic spiral.

This experiment suggests that leaf initiation can be accelerated by changes in the physical qualities of the shoot apical meristem. Turgor subjects the

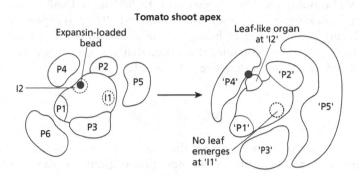

Fig. 6.12 The effects of applying expansin to the tomato shoot apex. Expansins increase the extensibility of the cell wall. Placing an expansin-loaded bead (red) onto the I2 site on the tomato shoot apex can induce the formation of a leaf-like organ at I2, and suppress leaf initiation at I1. This suggests that leaf initiation can be accelerated by changing the physical properties of the shoot apical meristem. (Adapted from Fleming *et al.*, 1977.)

outer walls of cells at the meristem surface to outward pressure. Therefore, exogenous expansin may cause the meristem to bulge by decreasing the ability of the outermost cell walls to resist outward pressure. It is unclear how such bulging accelerates leaf formation.

It has been proposed that the growth of existing leaf primordia may directly affect the pattern of physical forces in the meristem leading, directly or indirectly, to the bulge that is the next primordium. Two lines of evidence indicate that such a mechanism is unlikely to control phyllotaxy. Firstly, the surgical experiments discussed earlier in this case study suggest that the influence of existing primordia on future leaf initiation is chemical rather than physical. Secondly, some mutations dramatically alter primordium shape, and presumably the physical influences of primordium growth, without changing phyllotaxy. The *leafbladeless1* (*lbl1*) mutation of maize (case study 5.1) provides a dramatic example. Wild-type maize leaf primordia are key-ring-shaped, encircling the meristem. Extreme *lbl1* primordia are narrow crescents from which thread-like leaves develop. In both cases, phyllotaxy is alternate.

Timing of leaf specification

Surgical experiments on the shoot apical meristem suggest that the position of the next leaf primordium remains flexible until shortly before the primordium appears. The length of time between the initiation of successive nodes on the shoot is called a plastochron. Incisions in the apical meristem made more than half a plastochron before I1 emerges can shift I1 to a new position. Incisions made less than half a plastochron before I1's emergence cannot alter the location of I1. Therefore, the position of a new leaf primordium becomes fixed about half a plastochron before the primordium forms. However, leaf fate is specified much earlier than I1, as demonstrated by the expression of leaf-specific genes such as *YABBY* and *PHANTASTICA*

(*PHAN*) homologues. The expression of *YABBY* homologues is detected at I2 in the *Arabidopsis* shoot apical meristem, and *PHAN* is expressed as early as I4 during the initiation of bracts on the *Antirrhinum* inflorescence (see case study 5.1). It is probable, therefore, that the site of leaf initiation is selected before I2, but is only fixed during I1.

Initiation of phyllotaxy

In the early stages of shoot growth, there are too few existing leaf primordia to define a precise location for the next primordium. In plants with spiral phyllotaxy, this uncertainty is revealed by the large number of species in which leaves have an equal probability of spiralling clockwise or anticlockwise up the stem. Clover (*Trifolium* spp.) is a good example. The first two true leaves on a clover shoot form at right angles to the cotyledons. Spiral phyllotaxy begins with the initiation of the third leaf. Imagine the cotyledons and the first two true leaves arrayed on a clock face with the cotyledons pointing to 'twelve' and 'six', the first leaf pointing to 'nine', and the second leaf pointing to 'three' (Fig. 6.13). In many clovers, the third leaf has an equal probability of initiating at either of two positions. If the third leaf develops at 'half past seven', then subsequent leaves will form a clockwise spiral up the shoot. However, if the third leaf forms at 'half past ten', subsequent leaves will spiral anticlockwise. The selection of one of the two possible sites for leaf three requires a symmetry-breaking process. The nature of this

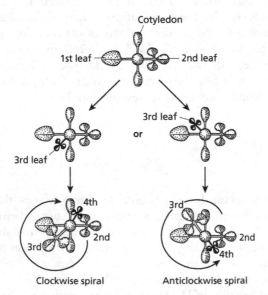

Fig. 6.13 The initiation of phyllotaxy in clover (*Trifolium* spp.). The first and second true leaves of the clover shoot are initiated perpendicular to the cotyledons (top). Spiral phyllotaxy begins with the initiation of the third leaf, the position of which determines whether leaves will spiral clockwise or anticlockwise up the shoot (middle and bottom). (Adapted from Rijven, 1968.)

is unknown but it has been suggested that a Turing mechanism may operate (see case study 6.1).

Case study 6.4: Coordination of leaf and vascular development

Leaf primordia must develop in association with vascular connections to the rest of the plant. The differentiation of the vascular connection is likely to involve an inductive signal emanating from the leaf primordium. Young leaves are thought to be major exporters of the hormone auxin, and auxin has been shown to stimulate vascular differentiation. Therefore, a reasonable hypothesis is that auxin synthesized by the leaf primordia induces differentiation of vascular connections to the stem. This case study considers the evidence for the stimulation of vascular development by leaf-derived auxin.

Induction of vascular development by auxin

As described in Chapter 1, vascular differentiation in the stem occurs in a continual acropetal wave. Vascular bundles in the stem merge with strands of procambium at the shoot apex. Cells at the base of the procambial strands differentiate into mature vascular tissue, and the strands extend acropetally as the shoot develops. At the shoot apical meristem, each new leaf primordium is associated with one or more strands of procambium in the underlying tissue. These strands extend into the leaf as it develops, providing the vascular connection between the leaf and the rest of the plant (Fig. 6.14).

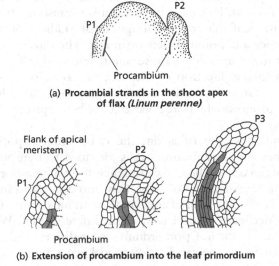

(a) Procambial strands in the shoot apex of flax (Linum perenne)

(b) Extension of procambium into the leaf primordium

Fig. 6.14 The development of the vascular connection to the leaf. (a) The apex of flax (*Linum perenne*) showing strands of procambium (red) associated with the youngest two leaf primordia. (b) Extension of procambium into a leaf primordium of flax. (Adapted from Waring & Phillips, 1970.)

The relationship between the leaf, auxin and vascular connections to the stem has been shown by simple surgical experiments. In *Coleus*, for example, removing a leaf primordium (either P1 or P2) prevents further differentiation of the strand of procambium linking that primordium to the stem. However, if the excised primordium is replaced by applied auxin, the normal pattern of vascular development continues.

Similar results are observed in experiments on vascular regeneration. When a wound severs a mature vascular bundle in the stem, a new vascular strand will differentiate from stem parenchyma until the bundle is repaired. Vascular regeneration is severely reduced when leaves above a wound are removed but can be restored if the excised leaves are replaced with auxin. As discussed in Chapter 4, auxin is transported basipetally through the shoot (i.e. from tip to base). Cutting off leaves below the wound does not prevent vascular strands from regenerating.

The canalization hypothesis

How then does auxin leaving the leaf induce a discrete vascular connection to existing procambial strands in the stem? A clue comes from the observation that the vascular system is the principal site of basipetal auxin transport in the shoot: polar auxin transport occurs primarily in elongated parenchyma cells associated with vascular bundles (see Fig. 4.5).

The close connection between auxin supply, vascular development and auxin transport has led to the **canalization hypothesis** in which a positive feedback mechanism links vascular differentiation to polar auxin transport (Fig. 6.15). Consider a group of undifferentiated cells beneath a leaf primordium. As the primordium exports auxin, auxin concentration rises immediately under the leaf. However, basipetal auxin transport through vascular tissue below the leaf will create a longitudinal gradient of auxin concentration and hence a diffusion-driven auxin flux. The canalization hypothesis proposes that such auxin fluxes cause undifferentiated cells to begin transporting auxin in the direction of the flux, i.e. to become polarized with respect to auxin transport. Furthermore, it is proposed that cells transporting auxin become progressively more polarized and hence progressively better at auxin transport.

Given a limited supply of auxin, the canalization hypothesis predicts competition between neighbouring cells for auxin to transport. Stochastic variations between cells in the initial efficiency of auxin transport will result in the channelling (canalization) of auxin flow into discrete strands of increasingly polarized cells. The level of auxin in these strands eventually rises above the threshold needed to stimulate vascular differentiation. With a discrete supply of auxin, i.e. the leaf primordium, and a discrete sink for auxin, i.e. existing vascular tissue, the canalization model generates strands of procambium that connect the two.

The canalization hypothesis predicts that if the efficiency of auxin transport is reduced, the vascular network will resemble a blocked drainage system, with local levels of auxin rising to give unusually thick strands of

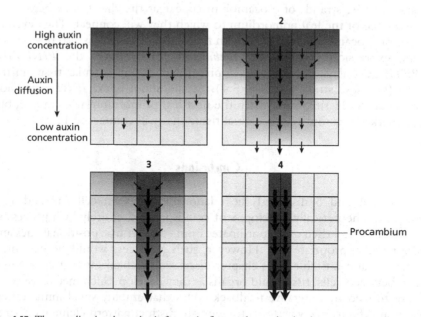

Fig. 6.15 The canalization hypothesis for auxin flow and vascular development in leaves. Auxin levels are shown in red, and polar auxin transport is indicated by arrows. 1, Cells exposed to a gradient of auxin begin net auxin transport in the direction of auxin diffusion. 2, 3, Positive feedback results in the channelling of auxin into an increasingly narrow strand of increasingly polarized cells. 4, Auxin concentration in the strand rises above the threshold required for differentiation into procambium.

procambium, impairing the continuity of the vascular system. These changes are observed in plants treated with auxin transport inhibitors and in some mutants with reduced auxin transport. Furthermore, treating *Arabidopsis* with auxin transport inhibitors shows that the vascular connection between the leaf and the stem is lost even at low inhibitor concentrations.

Signalling from the procambium to the leaf

It is probable that leaf primordia induce the differentiation of their vascular connections to the stem. However, the development of vascular tissue can occur in the absence of leaves. As described in case study 6.3, plants treated with auxin transport inhibitors and mutants with reduced auxin transport can develop pin-shaped apices in which leaf initiation is blocked. Vascular bundles continue to develop in such pins, albeit with altered morphology and pattern. This demonstrates that leaf-derived auxin is not necessary for vascular development at the shoot apex. Auxin exported from leaves, therefore, may function to modulate a semiautonomous pattern of vascular development in the stem.

It has also been suggested that signals from procambial strands in the shoot apex might promote leaf initiation, generating a feedback loop between procambial and leaf development. This suggestion arises because in some species, strands of procambium appear at the shoot apex before the emergence of the leaf primordium to which they will connect. The development of procambium associated with future leaf primordia is also revealed by gene expression patterns. In *Arabidopsis*, for example, the *PINHEAD* (*PNH*) gene is expressed in developing leaves and in vascular tissue in the stem (see case study 5.1, Fig. 5.4). At the shoot apex, *PNH* expression occurs at the I1 site, from which the next leaf primordium will emerge, but also marks presumptive procambial tissue that will connect to I2.

Conclusions

Chapters 4 and 5 discussed the relationship between cell fate and axial position. Theoretically, development could proceed entirely by a process in which cells read their axial coordinates from cell-extrinsic positional cues and adopt the appropriate fate. However, such a process would be extremely inflexible and unstable. This chapter discussed mechanisms that cross-reference between cells, tissues and organs as they develop. Such mechanisms are characterized by reiterative feedback cycles that amplify small initial differences and focus fuzzy patterns to generate distinct pattern elements. Cross-referencing mechanisms are particularly robust since variations in patterning are accommodated by appropriate changes in subsequent development.

Further reading

General

Animals

Greenwald, I. & Rubin, G.M. (1992) Making a difference: the role of cell–cell interactions in establishing separate identities for equivalent cells. *Cell* **68**, 271–281.

Plants

Hall, L.N. & Langdale, J.A. (1996) Molecular genetics of cellular differentiation in leaves. *New Phytologist* **132**, 533–553.

Case study 6.1: The pattern of trichomes on the *Arabidopsis* leaf

Reviews

The following review outlines the model described in this chapter:
Larkin, J.C., Marks, M.D., Nadeau, J. & Sack, F. (1997) Epidermal cell fate and patterning in leaves. *Plant Cell* **9**, 1109–1120.
For an alternative model see:
Szymanski, D.B., Lloyd, A.M. & Marks, M.D. (2000) Progress in the molecular genetic analysis of trichome initiation and morphogenesis in *Arabidopsis. Trends in Plant Science* **5**, 214–219.

Research papers

Larkin, J.C., Marks, M.D., Nadeau, J. & Sack, F. (1997) Epidermal cell fate and patterning in leaves. *Plant Cell* 9, 1109–1120.

Larkin, J.C., Oppenheimer, D.G., Lloyd, A.M., Paparozzi, E.T. & Marks, M.D. (1994) Roles of the *GLABROUS1* and *TRANSPARENT TESTA GLABRA* genes in *Arabidopsis* trichome development. *Plant Cell* 6, 1065–1076.

Larkin, J.C., Young, N., Prigge, M. & Marks, M.D. (1996) The control of trichome spacing and number in *Arabidopsis*. *Development* 122, 997–1005.

Schnittger, A., Folkers, U., Schwab, B., Jürgens, G. & Hülskamp, M. (1999) Generation of a spacing pattern: the role of *TRIPTYCHON* in trichome patterning in *Arabidopsis*. *Plant Cell* 11, 1105–1116.

Walker, A.R., Davison, P.A., Bolognesi-Winfield, A.C. *et al.* (1999) The *TRANSPARENT TESTA GLABRA1* locus, which regulates trichome differentiation and anthocyanin biosynthesis in *Arabidopsis*, encodes a WD40 repeat protein. *Plant Cell* 11, 1337–1349.

Turing mechanism

Turing, A.M. (1952) The chemical basis of morphogenesis. *Philosophical Transactions of the Royal Society of London, Series B* 641, 37–72.

Case study 6.2: The pattern of root hairs in *Arabidopsis*

Reviews

Dolan, L. (1996) Pattern in the root epidermis: an interplay of diffusible signals and cellular geometry. *Annals of Botany* 77, 547–553.

Scheres, B. (2000) Non-linear signalling for pattern formation? *Current Opinion in Plant Biology* 3, 412–417.

Scheres, B., McKhann, H.I. & van den Berg, C. (1996) Roots redefined: anatomical and genetic analysis of root development. *Plant Physiology* 111, 959–964.

Pattern of development in the root epidermis

Dolan, L., Duckett, C.M., Grierson, C. *et al.* (1994) Clonal relationships and cell patterning in the root epidermis of *Arabidopsis*. *Development* 120, 2465–2474.

Laser ablation experiments

Berger, F., Haseloff, J., Schiefelbein, J. & Dolan, L. (1998) Positional information in the root epidermis is defined during embryogenesis and acts in domains with strict boundaries. *Current Biology* 8, 421–430.

Genes that affect root hair development

Lee, M.M. & Schiefelbein, J. (1999) WEREWOLF, a MYB-related protein in *Arabidopsis*, is a position-dependent regulator of epidermal cell patterning. *Cell* 99, 473–483.

Masucci, J.D., Rerie, W.G., Foreman, D.R. *et al.* (1996) The homeobox gene *GLABRA 2* is required for position-dependent cell differentiation in the root epidermis of *Arabidopsis thaliana*. *Development* 122, 1253–1260.

Wada, T., Tachibana, T., Shimura, Y. & Okada, K. (1997) Epidermal cell differentiation in *Arabidopsis* determined by a *Myb* homolog, *CPC*. *Science* 277, 1113–1116.

Effects of ethylene signalling on root hair development

Cao, X.F., Linstead, P., Berger, F., Kieber, J. & Dolan, L. (1999) Differential ethylene sensitivity of epidermal cells is involved in the establishment of cell pattern in the *Arabidopsis* root. *Physiologia Plantarum* **106**, 311–317.

Masucci, J.D. & Schiefelbein, J.W. (1996) Hormones act downstream of *TTG* and *GL2* to promote root hair outgrowth during epidermis development in the *Arabidopsis* root. *Plant Cell* **8**, 1505–1517.

Patterning of hypocotyl stomata

Berger, F., Linstead, P., Dolan, L. & Haseloff, J. (1998) Stomata patterning on the hypocotyl of *Arabidopsis thaliana* is controlled by genes involved in the control of root epidermis patterning. *Developmental Biology* **194**, 226–234.

Hung, C-Y., Lin, Y., Zhang, M. *et al.* (1998) A common position-dependent mechanism controls cell-type patterning and *GLABRA 2* regulation in the root and hypocotyl epidermis of *Arabidopsis*. *Plant Physiology* **117**, 73–84.

Case study 6.3: Phyllotaxy

Surgical experiments and the field theory

Snow, M. & Snow, R. (1931) Experiments on phyllotaxis. I. The effect of isolating a primordium. *Philosophical Transactions of the Royal Society of London, Series B* **221**, 1–43.

Wardlaw, C.W. (1949) Experiments on organogenesis in ferns. *Growth* **13** (Suppl.), 93–131.

Both of the above are reviewed in:

Steeves, T.A. & Sussex, I.M. (1989) Organogenesis in the shoot: leaf origin and position. In: *Patterns in Plant Development*, 2nd edn, pp. 100–123. Cambridge University Press, Cambridge.

TERMINAL EAR1

Veit, B., Briggs, S.P., Schmidt, R.J., Yanofsky, M.F. & Hake, S. (1998) Regulation of leaf initiation by the *terminal ear 1* gene of maize. *Nature* **393**, 166–168.

Auxin

Reinhardt, D., Mandel, T. & Kuhlemeier, C. (2000) Auxin regulates the initiation and radial position of plant lateral organs. *Plant Cell* **12**, 507–518.

Expansin

Fleming, A.J., McQueen-Mason, S., Mandel, T. & Kuhlemeier, C. (1997) Induction of leaf primordia by the cell wall protein expansin. *Science* **276**, 1415–1418.

Reinhardt, D., Wittwer, F., Mandel, T. & Kuhlemeier, C. (1998) Localized upregulation of a new expansin gene predicts the site of leaf formation in the tomato meristem. *Plant Cell* **10**, 1427–1437.

Case study 6.4: Coordination of leaf and vascular development

Reviews

Aloni, R. (1987) Differentiation of vascular tissues. *Annual Review of Plant Physiology* **38**, 179–204.

Berleth, T., Mattsson, J. & Hardtke, C.S. (2000) Vascular continuity and auxin signals. *Trends in Plant Science* **5**, 387–393.

Nelson, T. & Dengler, N. (1997) Leaf vascular pattern formation. *Plant Cell* **9**, 1121–1135.

Coleus

Bruck, D.K. & Paolillo Jr, K.J. (1984) Replacement of leaf primordia with IAA in the induction of vascular differentiation in the stem of *Coleus*. *New Phytologist* **96**, 353–370.

The canalization hypothesis

Sachs, T. (1981) The control of the patterned differentiation of vascular tissues. *Advances in Botanical Research* **9**, 151–262.

Effects of auxin transport inhibitors

Mattsson, J., Sung, Z.R. & Berleth, T. (1999) Responses of plant vascular systems to auxin transport inhibition. *Development* **126**, 2979–2991.

CHAPTER 7

Light

This chapter and the next discuss the environmental regulation of development. Plants integrate a range of environmental cues to regulate morphology according to their habitat and to the time of year. Of these cues, light has the most profound effects on development. Plants are exquisitely sensitive to illumination, detecting the spectral quality, intensity, direction and duration of the light signal and modulating development in accordance with all of these factors. Developmental responses to light occur throughout post-embryonic growth, affecting processes ranging from germination to flowering. Before discussing specific responses, this chapter gives a brief introduction to light perception.

Light perception

The illumination of plants with different wavelengths of light shows that development is more responsive to some regions of the spectrum than to others. In particular, plants display high sensitivity to UV-B radiation (wavelengths from about 280 nm to 320 nm), UV-A radiation (from about 320 nm to 380 nm), blue light (from about 380 nm to 500 nm), red light (from about 620 nm to 700 nm) and far red light (from about 700 nm to 800 nm). It is these wavelengths to which plant **photoreceptors** are most sensitive. Plant photoreceptors are divided into four groups: **phytochromes**, which are primarily responsible for the perception of red and far red light; **cryptochromes** and **NPH1** (also called **phototropin**) which mediate responses to UV-A and blue wavelengths; and one or more unidentified UV-B receptors.

Phytochromes, cryptochromes and NPH1 all consist of proteins bound to light-absorbing pigments called **chromophores**. The spectral sensitivity of each photoreceptor depends on the ability of its chromophore(s) to absorb different wavelengths, i.e. on the chromophore's **absorption spectrum**. In response to light absorption, downstream signalling is mediated by the photoreceptor protein.

Phytochromes

Photoreversibility

Phytochromes were originally discovered because of the phenomenon of **photoreversibility**, in which responses promoted by a few minutes of dim red light are prevented by subsequent brief exposure to dim far red light. Photoreversibility occurs because phytochromes exist as two, interconvertible isomers, each of which has a characteristic absorption spectrum (Fig. 7.1a). One isomer, called **Pr**, absorbs most strongly in the red region of

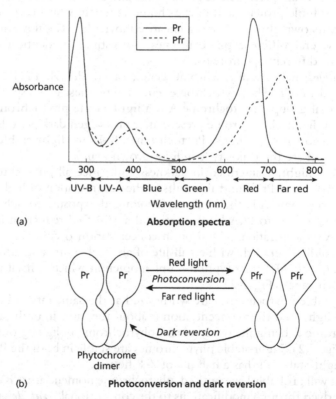

(a)   **Absorption spectra**

(b)   **Photoconversion and dark reversion**

Fig. 7.1 Light-absorption spectra and photoconversion of phytochromes. (a) Absorption spectra of Pr and Pfr. Both Pr and Pfr show some absorption at most wavelengths from UV-B to far red. In the red/far red region of the spectrum, Pr absorbs red wavelenths most strongly, whereas Pfr absorbs far red wavelengths most strongly. (b) Interconversion of Pr and Pfr. When Pr absorbs light it converts to Pfr (photoconversion) and when Pfr absorbs light it converts to Pr. Pfr also converts to Pr in a light-independent process known as dark reversion. Note that since phytochrome exists as a dimer, the transition between Pr and Pfr represents the transition between the Pr/Pr dimer and Pfr/Pfr dimer. It is possible that a transitory Pr/Pfr dimer is formed as an intermediate, and it has been suggested that such an intermediate may have biological activity. ((a) adapted from Chory, 1997.)

the spectrum (peak absorbance at about 660 nm). The other, called **Pfr**, absorbs best in the far red region (peak absorbance at around 730 nm). Light absorption by Pr leads to its conversion to Pfr (**photoconversion**). Similarly, light absorption by Pfr converts it to Pr. Because Pr and Pfr both absorb some light at all wavelengths, no light regime leads to the conversion of all phytochrome into just one form. However, phytochrome is synthesized as Pr and so seeds and seedlings grown in total darkness contain only this isomer. In addition to photoconversion, light-independent reversion of Pfr to Pr has been observed in many dicots (but not monocots). Although such reversion occurs in darkness and light, it is measured by the fall in Pfr concentration in the dark and hence is known as **dark reversion** (Fig. 7.1b).

Phytochrome types and nomenclature

Phytochromes are present in all green plants and at least some cyanobacteria. They are soluble proteins and exist as homodimers in which each subunit is bound to its own chromophore (a linear tetrapyrrole). The flowering plants contain several different phytochromes, consisting of identical chromophores but different apoproteins.

Arabidopsis has five phytochrome genes, called *PHYA*, *PHYB*, *PHYC*, *PHYD* and *PHYE*. The phytochromes encoded by these genes split into two fundamental groups. Phytochrome A is a **light-labile phytochrome** and is the predominant phytochrome present after prolonged dark periods in both imbibed seeds and seedlings. Phytochromes B–E are **light-stable phytochromes** and are the major type in light-grown plants.

PHYA is highly transcribed in darkness and, since all phytochromes are synthesized in the Pr form, this results in the accumulation of high levels of PrA (phytochrome A in the Pr form). Prolonged exposure to light and the conversion of PrA to PfrA leads to around a 100-fold reduction in phytochrome A concentration. The drop in concentration occurs partly because PfrA is rapidly degraded (with a half-life of about 1 hour) and also because light inhibits *PHYA* transcription—a phenomenon which is itself mediated by phytochromes (Fig. 7.2).

Light-stable phytochromes are synthesized at the same rate in both darkness and light, and their concentration is about the same in both cases, with phytochrome B being the predominant phytochrome in light-grown *Arabidopsis* (Fig. 7.2). Light-stable phytochromes are stable in both the Pr and Pfr forms (light-stable Pfr has a half-life of 7–8 hours).

As you will probably have noticed, phytochrome nomenclature is complex. It was derived through modifications to the conventional *Arabidopsis* genetic nomenclature (see p. 42). As with other *Arabidopsis* genes, the wild-type alleles of phytochrome genes are written in uppercase italics, *PHYA*, *PHYB*, etc. Also in line with the normal convention, the apoproteins encoded by the phytochrome genes are written in plain text uppercase, PHYA, PHYB, etc. However, the corresponding holoproteins (dimers complete with chromophores) are written phyA, phyB and so on. The Pr and Pfr forms of phyA are described as PrA and PfrA; those of phyB are written PrB and PfrB; etc. A similar system of nomenclature is applied to cryptochromes (see below).

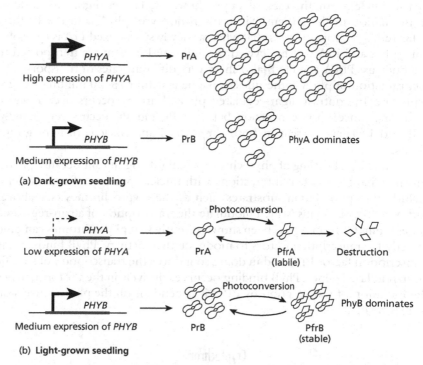

Fig. 7.2 The balance between phyA and phyB in dark- versus light-grown seedlings. (a) In dark-grown seedlings, the *PHYA* gene is highly expressed resulting in the accumulation of large amounts of phyA in the Pr form (PrA). The *PHYB* gene is expressed at a lower level. (b) The concentration of phyA declines rapidly when dark-grown seedlings are transferred to the light because of light-induced inhibition of transcription from the *PHYA* gene, and the rapid destruction of PfrA. In contrast, the *PHYB* gene is transcribed at a similar rate in both light and dark conditions, and PfrB is stable. Consequently, whereas phyA dominates in dark-grown seedlings, phyB dominates in light-grown seedlings.

Phytochrome kinase activity and phytochrome movement

In the Pfr form, phytochromes display kinase activity, undergoing autophosphorylation and also catalysing the phosphorylation of other proteins. It is probable that Pfr autophosphorylation modulates the interaction between phytochromes and other signalling molecules. Similarly, the phosphorylation of other proteins by Pfr is likely to change their activities. Interestingly, the targets for phytochrome kinase activity include the cryptochromes, suggesting that red/far red and blue/UV-A signal transduction interact at the level of the photoreceptors.

Another consequence of light absorption by phytochromes is a change in subcellular location. As described above, phytochromes are synthesized in the Pr form. In darkness, newly synthesized Pr is exclusively cytosolic, however exposure to red light causes the movement of phytochrome from the cytosol to the nucleus. In the case of phyB, this movement is inhibited by a subsequent pulse of far red light, suggesting that only PfrB is transported to the nucleus. In the case of phyA, however, far red light also induces accumulation in the nucleus. Since the majority of phyA is in the Pr form in far red light, this suggests that whereas newly synthesized PrA is cytosolic, both PfrA and PrA that has cycled through the Pfr form are transported to the nucleus. PhyA and phyB also show very different rates of transport. The accumulation of phyA in the nucleus happens within 15–20 minutes of light exposure. In contrast, light-regulated phyB shuttling occurs over a period of hours. Since photoconversion between Pr and Pfr occurs very rapidly, PrB and PfrB are present in both the cytosol and nucleus of light-grown plants.

The nuclear shuttling of phytochromes is highly significant because phytochromes have important interactions with nuclear proteins. These proteins include phosphorylation substrates such as the cryptochromes (see above) and Aux/IAA proteins (which modulate the transcription of auxin-regulated genes). Furthermore, it has been shown recently that phytochromes can bind directly to transcription factors to modulate their activity. PhyB binds to the transcription factor PIF3 and in doing so induces the transcription of specific light-regulated genes. PhyB binding occurs exclusively in the Pfr form, hence the initiation of transcription by PIF3 is dependent on the photoconversion of PrB to PfrB.

Cryptochromes

The generic term 'cryptochrome' was coined for blue and UV photoreceptors to reflect both the difficulty of identifying the photoreceptors and the fact that such receptors were known to be important to 'cryptogams' (an old term for seedless plants). More recently, genetic research in *Arabidopsis* has demonstrated that most responses to blue and UV-A wavelengths are mediated by two related photoreceptors. Following tradition, these are now called cryptochrome 1 (cry1) and cryptochrome 2 (cry2). Cry1 and cry2 are encoded by the *CRY1* and *CRY2* genes, respectively. The cryptochromes are soluble, apparently nuclear, proteins that occur in all plant organs. Each contains two chromophores, thought to be a flavin and a pterin.

Mutant analysis indicates that cry1 and cry2 have distinct but overlapping functions. These are discussed in more detail in the case studies. In an interesting parallel to phytochrome A, the level of cry2 declines rapidly when seedlings grown in darkness are exposed to blue light. The decline appears to be the result of increased cry2 degradation rather than a fall in transcription from the *CRY2* gene. In contrast, cry1 is light-stable.

Gene sequence analysis indicates that the cryptochromes evolved from an ancestral photolyase, an enzyme that mediates light-induced DNA repair. Interestingly, a similar but independent process appears to have occurred during animal evolution, resulting in proteins similar to plant cryptochromes. Animal cryptochromes have been identified in *Drosophila* and mammals where they are involved (as they are in plants) in the light regulation of circadian rhythms.

NPH1

Arabidopsis has a blue/UV-A photoreceptor that appears to be dedicated to phototropism, i.e. bending in response to unidirectional light (discussed in case study 7.4). This photoreceptor is encoded by the *NONPHOTOTRO-PIC HYPOCOTYL1* (*NPH1*) gene which was identified in a screen for non-phototropic mutants. *NPH1* encodes a plasma-membrane-associated protein kinase which is thought to have two flavin chromophores. In its holoprotein form, NPH1 is sometimes called 'phototropin'.

Developmental responses to light

In angiosperms, gametophytic and embryonic development are largely insensitive to light. However, light regulates development at all other stages of the life cycle. As discussed in case study 7.1 (see below), light is among the factors that can trigger germination, increasing the chances of germination at or near the soil surface. If seeds germinate in continual darkness, seedlings adopt an **etiolated** morphology appropriate to growth beneath the soil. This is characterized by rapid elongation of the hypocotyl or epicotyl in a region below an apical hook, a mode of growth that protects the shoot apical meristem and brings it rapidly to the soil surface. On reaching the light at the soil surface, seedlings undergo **photomorphogenesis**, characterized by enhanced root development, reduced stem elongation, opening of the apical hook and active leaf initiation, expansion and greening. The control of etiolation and photomorphogenesis are considered in case study 7.2.

After reaching the light, plants can respond to its spectral quality, direction and duration. Case study 7.3 discusses the effects of vegetational shade on growth and development. The case study focuses on **shade escape**, a response characterized by increased stem elongation and reduced branching. Case study 7.4 considers **phototropism**: the bending of an organ in relation to the direction of illumination. Finally, case study 7.5 discusses the regulation of flowering by day-length (the **photoperiod**).

Case study 7.1: Light-induced germination

Germination occurs by massive cell elongation in the radicle and embryonic shoot, coordinated with the mobilization of food reserves in the seed (Chapter 1). In some species, germination is accompanied by hydrolysis of endosperm cell walls near the radicle tip, weakening the physical barrier to radicle emergence.

For *Arabidopsis* and many other angiosperms, germination is promoted by light. Treatment with single wavelengths and mutational analysis indicate that light-induced germination is largely mediated by phytochromes. This case study discusses the role of phytochromes in the germination of *Arabidopsis* seeds.

Phytochrome mutants

Batches of newly imbibed *Arabidopsis* seeds display some germination in darkness, however germination is promoted by low amounts (**fluences**) of red light and inhibited by low amounts of far red light. After several days in darkness, imbibed seeds become dramatically more sensitive to light: they will germinate in response to a broad spectrum of radiation, from UV-B to far red, provided at very low fluences.

These contrasting responses suggest that more than one photoreceptor controls germination. Support for this idea comes from analysis of *phyA* and *phyB* mutants (Fig. 7.3). In *phyB* mutants, germination in darkness or after treatment with low fluences of red light is substantially reduced compared to wild type. This suggests that phyB is largely responsible for mediating germination under these conditions. The induction of germination by red light, and its inhibition by far red light, suggest that phyB promotes germination when it is in the Pfr form. The reason some seeds germinate in darkness is unclear; perhaps embryos contain some PfrB due to phyB synthesis and photoconversion during embryogenesis. In dry seeds, phytochromes undergo neither photoconversion nor dark reversion.

phyA mutants are specifically impaired in the highly light-sensitive response which develops when imbibed seeds are kept in total darkness. This suggests that phyA mediates germination under these conditions. The extreme light sensitivity of imbibed seeds develops because of the accumulation of high quantities of phyA in the Pr form during prolonged darkness (see **phytochromes**, above). Under these conditions, germination is promoted even when a tiny fraction of the accumulated PrA is converted to PfrA. Since PrA displays some absorption over the whole spectrum, this allows exposure to a very low fluence of any wavelength to induce germination.

Germination of *phyB* mutants is still promoted to an extent by red light and inhibited by far red light. This suggests that one or more of the other light-stable phytochromes (phyC, phyD and phyE) also regulate germination.

The responses of *Arabidopsis* to different light regimes allow germination in a wide range of environments. The fact that a proportion of *Arabidopsis* seeds germinate in darkness means that some seeds will always germinate when temperature and soil moisture allow. The promotion of germination by red light (mediated largely by phyB) will enhance the germination of *Arabidopsis* seeds on the soil surface in sunlight (which has a high ratio of red to far red light).

The extremely light-sensitive response mediated by phyA will promote germination in other environments. Firstly, it provides a means of inducing germination in buried seeds after exposure to flashes of light during soil disturbance. Secondly, it may promote germination of seeds buried just below the soil surface. Lastly, it will promote the germination of seeds on the soil surface but beneath a heavy canopy. Leaves absorb red wavelengths much more efficiently than far red wavelengths. Consequently, light beneath a canopy has a low ratio of red to far red wavelengths, preventing the phyB- but not the phyA-mediated germination response. Responses to leaf shade are discussed in detail in case study 7.3.

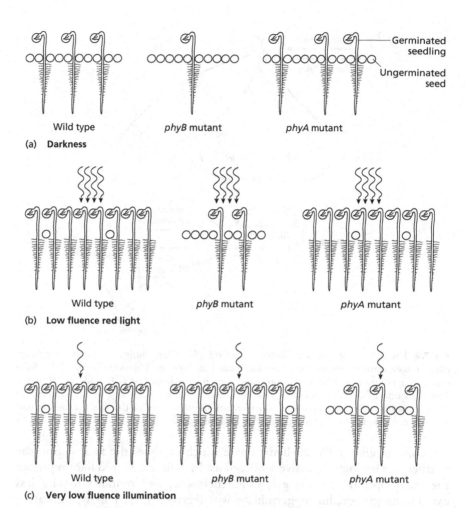

Fig. 7.3 Germination rates of seeds of wild-type and phytochrome mutants of *Arabidopsis* in different light environments. (a) Some wild-type seeds germinate even in darkness. This response is reduced in *phyB* mutants, suggesting that it requires phyB. (b) Germination of wild-type seeds is promoted by low fluence red light. This response is also reduced in *phyB* mutants, suggesting that it requires phyB. (c) When imbibed for several days in darkness, wild-type seeds become extremely light-sensitive and will germinate in response to very low fluence light across a broad spectrum. This response is lost in *phyA* mutants, suggesting that it requires phyA.

Gibberellins

The biosynthesis of gibberellins, a class of plant hormones, is required for germination in most species. For example, *Arabidopsis gibberellic acid1* (*ga1*) mutants are unable to synthesize gibberellins and fail to germinate even in the light unless gibberellins are provided exogenously. There is strong evidence that phytochromes mediate their effects on germination by influencing gibberellin biosynthesis and sensitivity (Fig. 7.4). Red light induces transcription of the *GA4* and *GA4H* genes of *Arabidopsis*. These encode

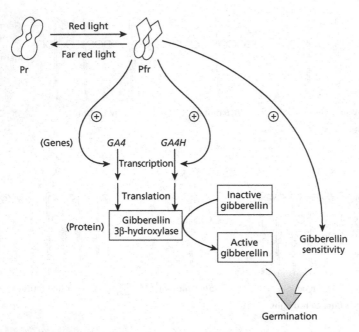

Fig. 7.4 The interaction between phytochrome and gibberellins during germination. Gibberellins induce germination in *Arabidopsis* and many other species. Phytochrome in the Pfr form increases gibberellin synthesis by promoting transcription from the *GA4* and *GA4H* genes. These encode gibberellin 3β-hydroxylase, which catalyses the final step in the synthesis of bioactive gibberellins. Pfr signalling also increases the sensitivity of the seed to gibberellins. Arrows accompanied by a 'plus' sign indicate positive regulation.

the enzyme gibberellin 3β-hydroxylase which catalyses the final step in the synthesis of biologically active gibberellins. In contrast, far red light represses the transcription of these genes. Furthermore, *ga1* mutants require less exogenous gibberellin to germinate when grown in red light, indicating that phytochrome signalling also increases the seed's sensitivity to gibberellins.

Case study 7.2: Seedling etiolation and photomorphogenesis

Angiosperm seedlings become **etiolated** in constant darkness (Fig. 7.5), a state characterized by reduced root development; a hook-shaped shoot; rapid stem or hypocotyl elongation; folded, unexpanded leaves and/or cotyledons; an inactive shoot apical meristem; and **etioplasts** (chloroplast precursors that lack chlorophyll). Etiolation is an adaptation to germination below the soil surface. Resources are directed towards rapid upward growth through the soil, with the hook-shaped shoot protecting the shoot apical meristem. The process of etiolation is sometimes called **skotomorphogenesis** (from the Greek 'skotos', meaning darkness).

Light-grown seedlings display **photomorphogenesis**, sometimes called de-etiolation (Fig. 7.5). This is characterized by extensive root development;

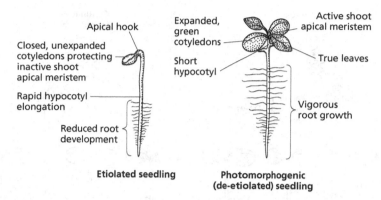

Fig. 7.5 Morphology of an etiolated and a photomorphogenic (de-etiolated) seedling.

slow stem elongation; straightening of the shoot; unfolding and expansion of leaves and cotyledons; the expression of genes necessary for pigment biosynthesis and the photosynthetic machinery; development of etioplasts into chloroplasts; and activation of the shoot apical meristem.

This case study discusses the perception of light by etiolated seedlings and the downstream events in the transition from etiolation to photomorphogenesis.

Light perception by the seedling

Photomorphogenesis can be induced by a broad spectrum of illumination, for example *Arabidopsis* seedlings de-etiolate in response to UV-A, blue, red or far red light. Some aspects of photomorphogenesis, such as changes in gene expression or the inhibition of hypocotyl elongation, can be induced by brief pulses of light. However, full de-etiolation requires continuous illumination.

The photoreceptors that mediate photomorphogenesis have been defined by mutant analysis. The phenotypes of *cry1* and *cry2* mutants of *Arabidopsis* indicate that both are required for full de-etiolation in response to UV-A or blue wavelengths (the *cry1* phenotype is shown in Fig 7.6). Consistent with the light lability of cry2, photomorphogenesis is mediated by the action of both cry1 and cry2 under low-intensity blue light, but largely by cry1 under high-intensity blue light. The phenotypes of phytochrome mutants (Fig. 7.6) show that phyB is needed for de-etiolation under continuous red light, while phyA mediates de-etiolation under continuous far red light. *phyB* mutants show residual photomorphogenesis under continuous red light, indicating that other light-stable phytochromes mediate de-etiolation.

Signal transduction downstream of these photoreceptors induces photomorphogenesis in two ways. Firstly, photosynthetic genes may be activated by direct positive regulation downstream of light perception. Secondly, light-induced signalling inactivates negative regulators of photomorphogenesis, and these are discussed below.

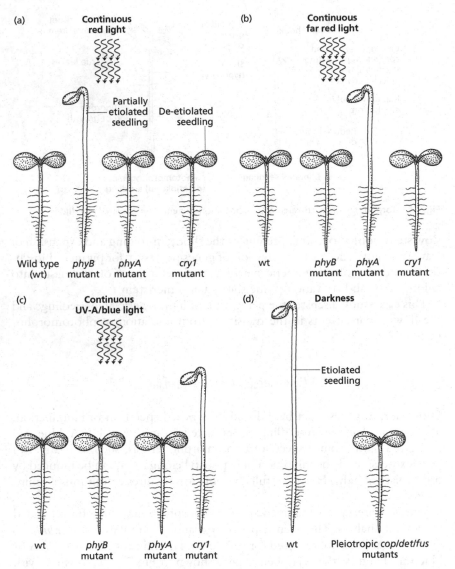

Fig. 7.6 Positive and negative regulation of photomorphogenesis in *Arabidopsis* revealed by mutants. Photomorphogenesis is promoted by continuous red light, continuous far red light and continuous UV-A/blue light in wild-type seedlings. The phenotypes of photoreceptor mutants indicate that the response to red light requires phyB (a); the response to far red light requires phyA (b); and the response to UV-A/blue light requires cry1 (c). (d) In darkness, wild-type seedlings are etiolated and photomorphogenesis is suppressed. The pleiotropic *cop/det/fus* mutants have a photomorphogenic phenotype in darkness, indicating that the *COP/DET/FUS* genes are required to suppress photomorphogenesis in dark-grown seedlings.

Negative regulators of photomorphogenesis

In evolutionary terms, photomorphogenesis is the ancestral pattern of development. Mosses, liverworts, ferns and some gymnosperms, notably conifers, have very similar patterns of development in light and darkness: etiolation is an adaptation largely restricted to angiosperm seedlings. It is therefore not surprising that genetic studies in *Arabidopsis* indicate that etiolation is a consequence of the inhibition of photomorphogenesis.

A number of *Arabidopsis* mutants have been identified that display photomorphogenesis when grown in continual darkness. This phenotype is described either as *de-etiolated* (*det*) or *constitutively photomorphogenic* (*cop*). *det* and *cop* mutations are recessive and result in a loss of gene function. The wild-type function of the *DET* and *COP* genes is, therefore, to suppress photomorphogenesis in dark-grown seedlings. Additional mutants with the *cop/det* phenotype were recovered in screens for purple seedlings and these mutants are called *fusca* (*fus*) (from the Latin 'fusca', meaning purple).

All the *cop/det/fus* mutants show some degree of photomorphogenesis in the dark. However, while some of the mutants show only a subset of the light-grown phenotype in darkness, 11 display a whole suite of photomorphogenic characteristics, i.e. the mutants are **pleiotropic** (Fig. 7.6). In darkness, pleiotropic *cop/det/fus* mutants have a short hypocotyl and open cotyledons, they express light-inducible genes and possess plastids intermediate between etioplasts and chloroplasts. Most of the 11 pleiotropic *cop/det/ fus* mutations were identified by more than one mutant screen and consequently the respective genes have more than one name. We will use the following 11 names: *COP1*, *COP8*, *COP9*, *COP10*, *COP16*, *DET1*, *FUS5*, *FUS6*, *FUS8*, *FUS11* and *FUS12* (after Wei & Deng, 1999).

The COP9 complex

Some of the *COP/DET/FUS* genes, including at least *COP8*, *COP9*, *FUS5* and *FUS6*, encode components of a multisubunit protein complex (Fig. 7.7). Because COP9 was the first member of the complex to be identified, this is known as the **COP9 complex** or the **COP9 signalosome**. The COP9 complex consists of eight different polypeptide subunits and is predominantly localized in the nucleus in both light- and dark-grown seedlings.

The COP9 complex probably has roles beyond the suppression of photomorphogenesis. Strong loss-of-function *cop/det/fus* alleles are seedling lethal, but plants carrying weaker alleles survive and show other abnormalities including early flowering, abnormal circadian rhythms and altered patterns of gene expression relating to pathogenesis and hypoxia (oxygen deficiency). Furthermore, the COP9 complex has been found in mice, *Drosophila* (the fruit fly) and budding yeast (*Schizosaccharomyces pombe*). By implication, the complex had, or has, an ancestral cellular function unrelated to plant development. To unify nomenclature across plants, animals and yeast, subunits of the COP9 complex are now designated CSN1–CSN8 (from <u>C</u>OP9 <u>s</u>ignalosome). (For details, see Deng *et al.*, 2000.)

Biochemically, the COP9 complex appears to be a multifunctional regulator of protein turnover. The complex is homologous to a subsection of the 26S proteasome, a much larger structure responsible for protein degradation, and acts as part of the intricate system that directs proteins to the

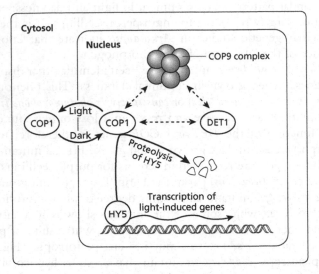

Fig. 7.7 Interactions between the COP9 complex, COP1, DET1 and HY5. The regulation of photomorphogenesis involves the shuttling of the COP1 protein between the cytoplasm and the nucleus. In darkness, COP1 is localized in the nucleus, where it accelerates the proteolysis of the HY5 transcription factor. In the light, COP1 is localized in the cytoplasm, where it cannot accelerate the destruction of HY5. HY5 promotes transcription from light-inducible genes. In darkness, therefore, nuclear COP1 indirectly prevents light-inducible gene expression. COP1 is constitutively cytoplasmic in *det1* mutants and in mutants that lack the COP9 complex, suggesting that interactions between COP1, the COP9 complex and DET1 are required to maintain COP1 in the nucleus in darkness. (Adapted from Torii & Deng, 1997.)

proteasome. However, the precise relationship between the COP9 complex and the proteasome is unclear.

COP1

Although many of the pleiotropic *cop/det/fus* mutations define proteins involved in the COP9 complex, some do not. Of these, the best understood is COP1. COP1 is a ubiquitin–protein ligase that acts by attaching ubiquitin to other proteins, labelling them for destruction. An important target for COP1-accelerated ubiquitination is the transcription factor, ELONGATED HYPOCOTYL5 (HY5) (Fig. 7.7). HY5 binds to a motif present in the promoters of many light-inducible genes and in doing so induces gene transcription. Loss-of-function *hy5* mutants have over-elongated hypocotyls in the light, demonstrating that *HY5* is required for photomorphogenesis. Presumably, therefore, HY5 promotes the transcription of genes that promote photomorphogenesis.

The activity of HY5, and hence the extent of photomorphogenesis, is regulated by at least three mechanisms. Firstly, the *HY5* gene is transcribed at a greater rate in the light than in the dark. Secondly, the HY5 protein is phosphorylated in darkness, causing a reduction in its activity. Thirdly, the HY5 protein has a shorter half-life when the plant is in the dark due to more rapid ubiquitination of HY5 by COP1. Interestingly, the phosphorylation of

HY5 in darkness makes the protein less senstive to COP1-mediated ubiquitination. It has been proposed that this prevents over-depletion of HY5 in the dark, allowing the seedling to respond more rapidly to light.

The regulation of COP1 activity by light occurs in part through changes in its cellular location. Immunolocalization and the use of COP1–reporter protein fusions show that COP1 is nuclear in darkness but cytoplasmic in the light. The stabilization of HY5 by light therefore involves the exclusion of COP1 from the nucleus (Fig. 7.7): in the cytoplasm, COP1 cannot target HY5, a nuclear protein, for destruction. COP1 is constitutively cytoplasmic in mutants that lack the COP9 complex, and also in *det1* mutants (*DET1* encodes a nuclear protein of unknown function). COP1 may, therefore, interact directly with the DET1 protein and the COP9 complex to maintain its nuclear location in the dark.

Other mechanisms must also regulate COP1 activity since the transition of COP1 from the nucleus to the cytoplasm after illumination occurs too slowly to be the means by which photomorphogenesis is initiated. The movement of COP1 may, therefore, be a mechanism for the maintenance rather than the initiation of photomorphogenesis.

Hormones

Several characteristics of etiolation and/or photomorphogenesis can be altered by hormone treatments and mutations affecting hormone synthesis and/or response. For example, two recessive mutations in *Arabidopsis*, *det2* and *constitutive photomorphogenesis and dwarfism* (*cpd*), result in similar, partially de-etiolated phenotypes in dark-grown seedlings. In darkness, *det2* and *cpc* mutants have short hypocotyls, no apical hook and express light-inducible genes. *DET2* and *CPD* encode enzymes that catalyse steps in the biosynthesis of the brassinosteroid hormone, brassinolide, and *det2* and *cpd* mutants have very low brassinolide content. Therefore, brassinolide is required for at least some aspects of etiolation.

Cytokinin, auxin and gibberellin have also been associated with the developmental changes that distinguish etiolation from photomorphogenesis. Addition of cytokinin or genetic perturbation of auxin signalling can cause partial de-etiolation in the dark. Gibberellin-deficient mutants are dwarfed relative to wild type, indicating that gibberellin is required for hypocotyl and stem elongation. As discussed in case study 7.1, gibberellin synthesis is necessary to induce germination in most species. It is possible that the increase in gibberellin concentration in germinating embryos allows the subsequent rapid upward growth of etiolated seedlings.

Case study 7.3: Shade escape

Plant development is regulated not only by the difference between darkness and light, but also by light quality, in particular the change in light quality due to shading by other plants. One of the main differences between sunlight

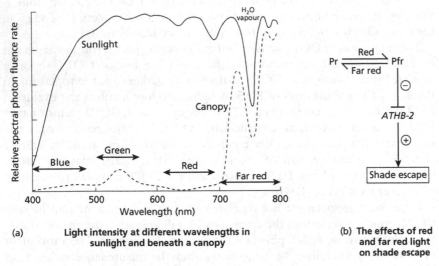

(a) **Light intensity at different wavelengths in sunlight and beneath a canopy**

(b) **The effects of red and far red light on shade escape**

Fig. 7.8 The ratio of red light to far red light regulates the shade escape response. (a) Comparing spectral fluence rates (light intensity at different wavelengths) in sunlight and canopy shade shows that the ratio of red light to far red light (R : FR) is much higher in sunlight. Therefore, R : FR is an indicator of the degree of shading by neighbouring plants. (b) If red is high relative to far red, as in sunlight, then Pr will convert to Pfr. High Pfr inhibits the transcription of the *ATHB-2* gene and hence inhibits shade escape. The line ending in a bar and accompanied by a 'minus' sign indicates negative regulation. The arrow accompanied by a 'plus' sign indicates positive regulation. ((a) from Smith, 1994 (Fig. 1) with kind permission from Kluwer Academic Publishers.)

and light that has passed through leaves is a fall in the ratio of red to far red wavelengths (R : FR) due to preferential absorption of red light by chlorophyll (Fig. 7.8). A drop in R : FR produces different responses in different plants. Most non-angiosperms are **shade tolerant** and respond by developing a morphology suited to shaded conditions, for example producing large but thin 'shade leaves'. In contrast, angiosperms usually respond with rapid internode and petiole elongation, a decrease in leaf area and thickness, enhanced apical dominance (i.e. reduced branching) and early flowering. This response is called **shade escape**.

This case study considers the influence of vegetation on R : FR, and R : FR signalling.

R : FR

R : FR is defined as the ratio in intensity of wavelengths from 655–665 nm (red) and 725–735 nm (far red). In open environments, the daytime R : FR is about 1.15 : 1 regardless of whether the sky is clear or overcast. Due to the presence of chlorophyll and other pigments, leaves strongly absorb wavelengths below 700 nm, but either transmit or reflect most light of wavelength 700–800 nm (Fig. 7.8a). Therefore, R : FR falls dramatically beneath a canopy due to the depletion of red wavelengths, reaching values as low as

0.05 : 1. R : FR also falls in the proximity of vegetation due to low R : FR in the light reflected from leaves. This allows some species to initiate **shade avoidance** responses before actual shading occurs by reacting to light reflected from neighbouring plants.

The importance of R : FR to shade escape has been demonstrated by supplying plants with a constant amount of photosynthetically active radiation (in the range 400–700 nm) but supplementing this with either red or far red light to generate different R : FR environments. In such conditions, shade escape responses are particularly pronounced because, unlike plants exposed to natural leaf shade, photosynthesis is not reduced.

Perception of R : FR

The absorbance peak of Pr occurs at a wavelength of about 660 nm while that of Pfr occurs at a wavelength of about 730 nm, making phytochrome exquisitely sensitive to R : FR. The function of phytochromes in shade escape has been demonstrated by the photoreversibility of some shade escape responses. For example, a pulse of red light decreases the rate of stem elongation, but this effect is negated by a subsequent pulse of far red light. It is estimated that in sunlight the photoconversion between Pr and Pfr equilibrates with about 55–60% of total phytochrome in the Pfr form. In contrast, under deep canopy shade, the proportion of Pfr can fall to less than 10% of the total phytochrome. This suggests that Pfr blocks the shade escape response (Fig. 7.8b).

phyA mutants of *Arabidopsis* display more or less normal shade escape, whereas *phyB* mutants display constitutive shade escape. When grown in white light, *phyB* mutants have elongated hypocotyls and petioles, increased apical dominance and flower early. Therefore, functional PfrB, but not PfrA, is required to inhibit shade escape. A small increase in shade escape responses still occurs in *phyB* mutants after lowering R : FR. For example, *phyB* mutants flower earlier than wild type under white light, but earlier still if R : FR is lowered. This suggests that other light-stable phytochromes also inhibit shade escape. Genetic analysis has implicated both phyD and phyE in the remaining responsiveness to R : FR of *phyB* mutants.

Signal transduction

In *Arabidopsis*, transcription from the *ATHB-2* gene is induced within 15 minutes by a drop in R : FR. *ATHB-2* encodes a homeodomain transcription factor and its repression in high R : FR is dependent on phyB and at least one other light-stable phytochrome. Constitutive expression of an *ATHB-2* transgene leads to constitutive shade escape morphology: plants have elongated hypocotyls and petioles, smaller leaves, a reduced root mass, and flower early. In contrast, plants with reduced *ATHB-2* expression due to the presence of an antisense *ATHB-2* transgene display the opposite phenotype. This suggests that the ATHB-2 transcription factor mediates downstream shade escape responses (Fig. 7.8b).

Case study 7.4: Phototropism

Plants change their direction of growth in response to changes in the direction of illumination. This bending response is called **phototropism**. Normally, shoots are positively phototropic and bend towards the light source, while roots are negatively phototropic and bend away. Leaves have more complex phototropic responses that affect both the position and orientation of the leaf lamina. In many species, if only part of the lamina is illuminated, the leaf petiole is induced to bend and/or twist in a way that moves the rest of the lamina out of the shade. This phenomenon allows the formation of **leaf mosaics** in which mutual shading between leaves in a canopy is minimized. In some species, the orientation of the lamina is also regulated according to light direction through bending at the junction between the lamina and the petiole. In very bright light, the lamina may be oriented parallel to incident light, reducing light exposure. In contrast, in dim light, laminas are often oriented perpendicular to the direction of illumination, enhancing light capture.

Phototropism in roots, stems and petioles normally occurs through differential cell expansion. For example, shoots bend towards a unilateral light source because cell elongation on the illuminated side of the stem decreases and cell elongation on the shaded side often (but not always) increases (shown for a grass coleoptile in Fig. 7.9). This rapid response can occur within 5 minutes of a transition from uniform to unilateral illumination. Changes in the orientation of the leaf lamina in response to unidirectional light are usually caused by reversible turgor changes in motor cells at the lamina–petiole junction.

Perception of light and signal transduction in phototropism

In most species, UV-A and blue wavelengths induce the greatest phototropic

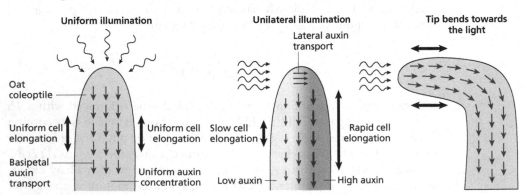

Fig. 7.9 The Cholodny–Went hypothesis of phototropism in oat coleoptiles. Unilateral illumination induces lateral auxin transport in the coleoptile tip leading to uneven auxin distribution in the coleoptile and to changes in cell expansion rates. The changes in expansion rates cause the coleoptile to bend towards the light.

response, suggesting that phototropism is primarily mediated by blue/UV-A photoreceptors. Consistent with this, *nph1* mutants of *Arabidopsis* show greatly reduced, but not zero, phototropism in blue light, indicating that NPH1 and one or more other blue/UV-A photoreceptors mediate the response.

A non-phototropic phenotype also results from loss-of-function mutations in the *NPH3* gene of *Arabidopsis*. The biochemical function of the NPH3 protein is unknown but it is likely to be the first component in a signal transduction pathway downstream of NPH1. NPH1 is a light-activated protein kinase that autophosphorylates after light absorption. The NPH3 protein contains predicted phosphorylation sites and has been shown to interact directly with NPH1. Therefore, it is likely the NPH3 is another target for NPH1 phosphorylation. The ROOT PHOTOTROPISM2 (RPT2) protein is similar in sequence to NPH3 and is also likely to function in signal transduction in phototropism since *rpt2* mutants have reduced phototropic responses in both the root and the shoot.

A mechanism by which unidirectional light might induce organ bending is set out in the **Cholodny–Went hypothesis**. This was formulated in the 1920s to explain the results of experiments on the oat coleoptile (the sheath that surrounds the shoot of grass seedlings).

The coleoptile tip is pivotal to phototropic responses. If the tip is removed, cell elongation in the coleoptile ceases, indicating that the tip produces a signal required for elongation. This signal can be collected by placing tips onto agar blocks and the amount of signal in the block can be assayed by placing it on a decapitated coleoptile and measuring elongation. If an isolated coleoptile tip is illuminated from the side, export of the elongation-promoting factor decreases from the illuminated half and increases in the shaded half. This change is prevented if the two halves of the tip are completely separated by a thin sheet of glass, suggesting that it results from the transport of the factor from the illuminated to the shaded side.

One of the substances produced by shoot tips that promotes cell elongation is auxin. The Cholodny–Went hypothesis proposes that perception of unidirectional light results in auxin redistribution in the tip with consequent changes in the concentration of auxin exported into the coleoptile and in cell elongation in the illuminated and shaded sides (Fig. 7.9).

Abundant physiological evidence from a variety of species supports a role for auxin redistribution in phototropism. For example, in some species, radiolabelled auxin applied to the tip of seedlings concentrates in the shaded tissue after exposure to unidirectional light. Also, phototropism is reduced in a dose-dependent manner by treatment with auxin transport inhibitors, consistent with those blocking auxin redistribution from the illuminated to the shaded tissue. However, changes in the concentration of endogenous auxin in the illuminated and shaded halves of stems cannot always be detected prior to phototropic bending. Therefore, although the Cholodny–Went hypothesis can be broadly applied, in its simplest form it does not fully explain phototropism. It is important to note that the Cholodny–Went hypothesis does not exclude differential cell expansion caused by changes in auxin sensitivity.

The requirement for auxin signalling in phototropism is further supported by the greatly reduced phototropic responses of some auxin signalling mutants, such as the *nph4* mutants of *Arabidopsis*. *NPH4* (also known as *ARF7*) encodes an Auxin Response Factor (ARF), a protein that regulates gene transcription in response to auxin. Reduction of both phototropism and auxin response in *nph4* hypocotyls is consistent with the idea that phototropism is mediated by auxin-induced changes in cell elongation. Interestingly, *nph4* seedlings also have a reduced gravitropic response (see case study 8.1).

Case study 7.5: Photoperiodic control of flowering

Day-length, the **photoperiod**, is an indicator of the season (except at the equator). The photoperiod regulates a wide range of developmental processes, including bud formation and dormancy, stem elongation, the formation of storage organs, vegetative reproduction, and flowering. The photoperiodic control of flowering ensures that flowers are produced in a favourable season and, furthermore, it allows floral synchronization in local populations, giving more efficient cross-pollination.

This case study considers how the photoperiod is measured, and discusses the transduction of the photoperiodic signal.

Photoperiod requirements

Plants are divided into three main groups depending on their photoperiod requirement for flowering. In **long-day plants** (LDPs) such as *Arabidopsis*, flowering is promoted by photoperiods above a critical length. In **short-day plants** (SDPs) such as soybean (*Glycine max*), flowering is promoted by photoperiods below a critical length. In **day-neutral plants**, such as most varieties of tobacco (*Nicotiana tabacum*), the photoperiod does not directly influence flowering.

LDPs and SDPs are subdivided further according to whether or not the photoperiodic requirement is absolute. Flowering in *Arabidopsis* is promoted by long days but eventually occurs in short days, making *Arabidopsis* a **facultative** LDP. Soybean, however, only flowers after exposure to short photoperiods and is described as an **obligate** SDP.

The responses of some plants are too complex to fit into any of the main groups. Fat hen (*Chenopodium album*), for example, only flowers when days are of intermediate length, remaining vegetative when days are either particularly long or short.

Measuring the photoperiod

Counterintuitively, photoperiodic responses are controlled primarily by night-length. Hence flowering in SDPs is promoted by nights longer than a critical length, while flowering in LDPs is promoted by nights shorter than a critical length. The importance of night-length was established by

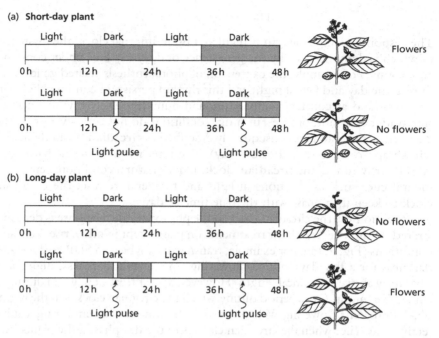

Fig. 7.10 The effects of night-time light treatments on photoperiodic flowering. Whereas day-time dark treatments have little effect on flowering, interrupting nights with brief periods of illumination can inhibit flowering in short-day plants (a) and can promote flowering in long-day plants (b).

studying the effects of night-time light treatments. Brief night-time illumination prevents flowering of SDPs in short photoperiods, but promotes flowering in LDPs under the same conditions (Fig. 7.10). In both cases, plants respond to an interrupted long night as if exposed to two short nights. In contrast, interrupting days with short periods of darkness has little or no effect on flowering.

The effects of exposing only particular sections of the shoot to inductive photoperiods suggest that perception of night-length occurs in mature leaves. In some species exposure of a single leaf to the correct photoperiod induces flowering. Furthermore, leaves cut from a plant that has been induced to flower and grafted onto a plant kept in non-inductive conditions often induce flowering in the recipient. This suggests that inductive photoperiods stimulate mature leaves to produce a flower-promoting signal (or combination of signals), which historically has been called **florigen**.

Whilst in some plants, the photoperiodic response is mediated by a positive signal, in others, such as in many pea varieties, a negative signal is required. These peas are LDPs and grafting experiments indicate that, in short days only, cotyledons and leaves produce a floral inhibitor. This inhibitor acts to overcome a constitutively produced promotor of flowering synthesized in leaves and in roots and/or cotyledons.

The circadian clock

The response to night-length is regulated by an internal clock, the existence of which is demonstrated by the persistence of daily rhythms under constant conditions. For example, the expression of photosynthesis-related genes rises during the day and falls at night and this rhythm persists in continual light or darkness, and at constant temperature and humidity. Daily rhythms occurring under constant, or **free-running**, conditions do not cycle over precisely 24-hour periods and are consequently described as **circadian** (from the Latin 'circa': approximately, and 'dies': day). The internal mechanism that regulates the rhythms is the **circadian clock**. Under natural conditions, environmental cues such as variations in light and temperature reset the circadian clock to keep it in phase with the true time of day.

A function for the circadian clock in the photoperiodic response is demonstrated by the effects of light treatments on plants kept in otherwise constant conditions (Fig. 7.11). For example, transferring soybean, a SDP, to constant darkness for a few days induces flowering. In constant darkness, nights are obviously above the critical length. However, a light treatment given during a subjective 'night' (i.e. a period during which the circadian clock is in the night phase) prevents flowering. When the same treatment is given during a subjective 'day' (i.e. when the circadian clock is in the day phase), the plants still flower. In constant darkness, therefore, the circadian clock determines when plants respond to 'night-time' light treatments. Similar experiments show that the circadian clock has the same function in LDPs.

This result is supported and extended by analysis of flowering in mutants

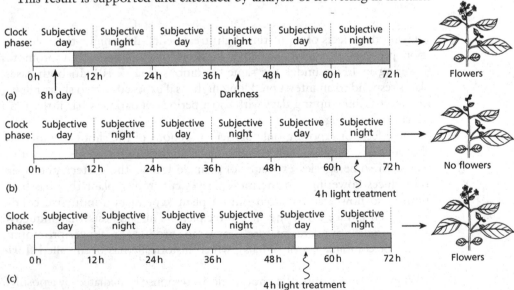

Fig. 7.11 Regulation of the photoperiodic response to light treatments by the circadian clock in soybean. Soybean is a SDP that requires several long nights to induce flowering. (a) The plant can be induced to flower if an 8-hour day is followed by 64 hours of darkness. The induction is prevented if a 4-hour light treatment is given during a subjective 'night' (b), but not if the light treatment is given during a subjective 'day' (c). (Data from Salisbury & Ross, 1992.)

with disrupted clock function. For example, the *TIMING OF CAB EXPRESS-ION1 (TOC1)* gene of *Arabidopsis* encodes a putative transcription factor that is thought to be a component of the circadian clock. *toc1* mutants have shortened circadian rhythms of around 21 hours under free-running conditions and have disrupted responses to conventional, 24-hour light–dark cycles. For example, wild-type plants flower late under short-day cycles of 7 hours light, 17 hours dark (7L : 17D) or 8 hours light, 16 hours dark (8L : 16D), and flower early under long-day cycles of 16 hours light, 8 hours dark (16L : 8D). In contrast, *toc1* mutants flower early under 7L : 17D but late under 8L : 16D and 16L : 8D. *toc1* mutants display the appropriate photoperiodic response if given 21-hour light–dark cycles, flowering late in short days given in the form 7L : 14D and early in long days of 14L : 7D. This suggests that the length of the day–night cycle must coincide with the period of the circadian clock for a correct photoperiodic response.

It has been proposed that the circadian clock may modulate the transduction of light signals so that light perception only induces a photoperiodic response when the clock is in the night phase (Fig. 7.12). This mechanism is called **gating**. When the clock is in the day phase, signal transduction between light and photoperiodic flowering responses is blocked. During the night phase, the metaphorical gate is opened and there is no block to signal transduction. Hence light perceived when the clock is in the night phase can prevent flowering in SDPs and promote flowering in LDPs. In other words, it may be that plants do not measure the length of the night at all. Instead,

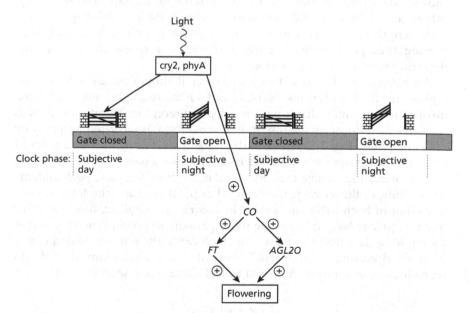

Fig. 7.12 A model for the photoperiodic regulation of flowering in *Arabidopsis*. The photoperiodic response is induced by light absorbed by cry2 and phyA. When the circadian clock is in the day phase, signal transduction between these photoreceptors and *CO* is blocked. When the clock is in the night phase, signal transduction is allowed, increasing *CO* expression and inducing flowering. Hence flowering is induced when night length is shorter than the period for which the gate is open. Arrows accompanied by a 'plus' sign indicate positive regulation.

they may determine when a critical night length has been reached by comparing the duration of the dark period with the period for which the gate is
open. Light perceived when the gate is open would indicate a short night.

The gating mechanism may require the gene *EARLY FLOWERING3*
(*ELF3*). *elf3* mutants of *Arabidopsis* flower early in both short and long days,
suggesting that photoperiodic promotion of flowering is constitutively active
in the mutants. The mutants also fail to gate the light activation of *CAB* genes
(which encode chlorophyll-binding proteins). In wild-type plants transferred
to continual darkness, brief light treatments increase *CAB* gene expression
when the clock is in the day phase but not when the clock is in the night phase.
Therefore, the induction of *CAB* expression by light is gated with the gate
open during the day phase and closed during the night phase. In *elf3* mutants,
however, light treatments increase *CAB* expression during both subjective
days and nights, indicating that *ELF3* is required to close the gate. If *elf3*
mutants are unable to close the gate to the photoperiodic response during the
day phase, then flowering will be induced by light perceived throughout the
day, causing early flowering irrespective of day length. Therefore, the early
flowering phenotype of *elf3* mutants most probably is a secondary consequence of the gating defect.

Photoreceptors mediating the photoperiodic response

In many SDPs, the inhibition of flowering by night-time light treatments
shows red/far red reversibility. A pulse of red light inhibits flowering and a
subsequent pulse of far red light removes the inhibition. This suggests that
night-length measurement in SDPs relies primarily on phytochromes. Supporting this, *phyB* mutants of the SDP *Sorghum bicolor* are day neutral,
flowering early under short and long days.

Arabidopsis, a LDP, is induced to flower if short days are followed by
night-time light treatments of blue, red or far red light, suggesting the
involvement of more than one class of photoreceptor. Mutational analysis
indicates the photoperiodic response of *Arabidopsis* is mediated primarily by
cry2 and phyA (Fig. 7.12) since both *cry2* and *phyA* mutants have a reduced
photoperiodic response and flower late during long days.

It is interesting to note that, like *phyB* mutants of *Sorghum*, *phyB* mutants
of *Arabidopsis* flower early, indicating that phyB mediates the inhibition of
flowering in both SDPs and LDPs. In contrast to *Sorghum*, however, *phyB*
mutants of *Arabidopsis* retain the photoperiodic response, flowering earlier
during long days than short days. In *Arabidopsis*, the phyB-mediated inhibition of flowering is considered primarily as a mechanism that delays
reproduction in sunlight compared to leaf shade (see case study 7.3).

CONSTANS

The CONSTANS (CO) transcription factor is an important downstream
regulator of the photoperiodic signal (Fig. 7.12). Long-day promotion of
flowering is absent in *co* mutants of *Arabidopsis*, demonstrating that *CO* is
required for photoperiodism. The *CO* gene is expressed in leaves and the

shoot apex. On the transition from short to long days, the level of *CO* mRNA increases by about threefold. This increase is severely curtailed in *cry2* mutants, which have a reduced photoperiodic response (see above). Long-day induction of flowering, therefore, is probably mediated by an increase in the concentration of the CO transcription factor. Supporting this conclusion, constitutive expression of a *CO* transgene induces early flowering even in short days.

Direct targets for transcriptional activation by CO include the genes *FLOWERING LOCUS T* (*FT*) and *AGAMOUS-LIKE 20* (*AGL20*, also called *SOC1*), both of which promote flowering. *AGL20* encodes a MADS domain transcription factor similar to AGAMOUS (see case study 5.4). *FT* encodes a protein with similarities to RAF kinase inhibitor protein (RKIP), which regulates the activity of the RAF intracellular kinase signalling cascade. It is at this point that the long-day induction of flowering converges with other mechanisms that control flowering in *Arabidopsis*: in addition to *CO*, *FT* and *AGL20* are regulated by vernalization and by an autonomous floral promotion pathway (see case studies 8.4 and 9.3, respectively).

Conclusions

Light provides the energy for photosynthesis and therefore it is not surprising that plant development is extremely responsive to the light environment. The wavelengths most efficiently absorbed for photosynthesis are those in the blue and red regions of the spectrum and these regions are highly efficient at inducing developmental responses. Responses to blue light are mediated by cryptochromes and NPH1, and those to red light are mediated by phytochromes. Phytochromes also mediate responses to far red wavelengths. In contrast to blue and red light, far red light is not absorbed for photosynthesis, making the proportion of far red wavelengths in light an indicator of vegetational shade.

Light provides an enormous amount of information and plants respond to the direction, intensity, spectral quality and duration of the light signal. Although this chapter has considered each of these aspects of light individually, many developmental processes are controlled by more than one light cue. For example, flowering in *Arabidopsis* is promoted by long days but this response is modulated by light quality, occurring earlier under leaf shade. Therefore, the ability of plants to adapt their growth and development to the prevailing light conditions involves substantial integration of the different classes of information provided by light.

Further reading

General

Chory, J. (1997) Light modulation of vegetative development. *Plant Cell* 9, 1225–1234.
Kendrick, R.E. & Kronenberg, G.H.M. (eds) (1994) *Photomorphogenesis in Plants*, 2nd edn. Kluwer Academic Publishers, Dordrecht.

Von Arnim, A. & Deng, X-W. (1996) Light control of seedling development. *Annual Review of Plant Physiology and Plant Molecular Biology* **47**, 215–244.

Photoreceptors

Phytochromes

Nagy, F. & Schäfer, E. (2000) Nuclear and cytosolic events of light-induced phytochrome-related signaling in higher plants. *EMBO Journal* **19**, 157–163.
Neff, M.M., Fankhauser, C. & Chory, J. (2000) Light: an indicator of time and place. *Genes and Development* **14**, 257–271.

Cryptochromes

Cashmore, A.R., Jarillo, J.A., Wu, Y-J. & Liu, D. (1999) Cryptochromes: blue light receptors for plants and animals. *Science* **284**, 760–765.
Lin, C. (2000) Plant blue-light receptors. *Trends in Plant Science* **5**, 337–342.

NPH1

Christie, J.M., Reymond, P., Powell, G.K. *et al.* (1998) *Arabidopsis* NPH1: a flavoprotein with the properties of a photoreceptor for phototropism. *Science* **282**, 1698–1701.
Lin, C. (2000) Plant blue-light receptors. *Trends in Plant Science* **5**, 337–342.

Case study 7.1: Light-induced germination

Phytochrome mutants

Reed, J.W., Nagatani, A., Elich, T.D., Fagan, M. & Chory, J. (1994) Phytochrome A and phytochrome B have overlapping but distinct functions in *Arabidopsis* development. *Plant Physiology* **104**, 1139–1149.
Whitelam, G.C. & Devlin, P.F. (1997) Roles of different phytochromes in *Arabidopsis* photomorphogenesis. *Plant, Cell and Environment* **20**, 752–758.

Gibberellins

Karssen, C.M., Zagórski, S., Kepczynski, J. & Groot, S.P.C. (1989) Key role for endogenous gibberellins in the control of seed germination. *Annals of Botany* **63**, 71–80.
Yamaguchi, S., Smith, M.W., Brown, R.G.S., Kamiya, Y. & Sun, T. (1998) Phytochrome regulation and differential expression of gibberellin 3β-hydrolase genes in germinating *Arabidopsis* seeds. *Plant Cell* **10**, 2115–2126.

Case study 7.2: Seedling etiolation and photomorphogenesis

Light perception

Lin, C. (2000) Plant blue-light receptors. *Trends in Plant Science* **5**, 337–342.
Neff, M.M., Fankhauser, C. & Chory, J. (2000) Light: an indicator of time and place. *Genes and Development* **14**, 257–271.

COP/DET/FUS genes

Chamovitz, D.A. & Deng, X-W. (1997) The COP9 complex: a link between photomorphogenesis and general developmental regulation? *Plant, Cell and Environment* **20**, 734–739.
Deng., X-W., Dubiel, W., Wei, N. *et al.* (2000) Unified nomenclature for the COP9 signalosome and its subunits: an essential regulator of development. *Trends in Genetics* **16**, 202–203.

Torii, K.U. & Deng, X-W. (1997) The role of COP1 in light control of *Arabidopsis* seedling development. *Plant, Cell and Environment* **20**, 728–733.

Wei, N. & Deng, X-W. (1996) The role of the *COP/DET/FUS* genes in light control of *Arabidopsis* seedling development. *Plant Physiology* **112**, 871–878.

Wei, N. & Deng, X-W. (1999) Making sense of the COP9 signalosome. *Trends in Genetics* **15**, 98–103.

HY5

Chattopadhyay, S., Ang, L-H., Puente, P., Deng, X-W. & Wei, N. (1998) *Arabidopsis* bZIP protein HY5 directly interacts with light-responsive promoters in mediating light control of gene expression. *Plant Cell* **10**, 673–683.

Hardtke, C.S., Gohda, K., Osterlund, M.T. *et al.* (2000) HY5 stability and activity in *Arabidopsis* is regulated by phosphorylation in its COP1 binding domain. *EMBO Journal* **19**, 4997–5006.

Jarillo, J.A. & Cashmore, A.R. (1998) Enlightenment of the COP1–HY5 complex in photomorphogenesis. *Trends in Plant Science* **3**, 161–163.

Osterlund, M.T., Hardtke, C.S., Wei, N. & Deng, X-W. (2000) Targeted destabilization of HY5 during light-regulated development of *Arabidopsis*. *Nature* **405**, 462–466.

Brassinosteroids

Clouse, S.D. (1996) Molecular genetic studies confirm the role of brassinosteroids in plant growth and development. *Plant Journal* **10**, 1–8.

Li, J. & Chory, J. (1999) Brassinosteroid actions in plants. *Journal of Experimental Botany* **50**, 275–282.

Case study 7.3: Shade escape

General reviews

Smith, H. (1994) Sensing the light environment: the functions of the phytochrome family. In: *Photomorphogenesis in Plants*, 2nd edn (ed. R.E. Kendrick & G.H.M. Kronenberg), pp. 377–416. Kluwer Academic Publishers, Dordrecht.

Smith, H. & Whitelam, G.C. (1997) The shade avoidance syndrome: multiple responses mediated by multiple phytochromes. *Plant, Cell and Environment* **20**, 840–844.

ATHB-2

Steindler, C., Matteucci, A., Sessa, G. *et al.* (1999) Shade avoidance responses are mediated by the ATHB-2 HD-Zip protein, a negative regulator of gene expression. *Development* **126**, 4235–4245.

Case study 7.4: Phototropism

General reviews

Firn, R.D. (1994) Phototropism. In: *Photomorphogenesis in Plants*, 2nd edn (ed. R.E. Kendrick & G.H.M. Kronenberg), pp. 659–681. Kluwer Academic Publishers, Dordrecht.

Iino, M. (1990) Phototropism: mechanisms and ecological implications. *Plant, Cell and Environment* **13**, 633–650.

NPH1

See references under 'photoreceptors', above. Also recommended is:

Liscum, E. & Briggs, W.R. (1995) Mutations in the *NPH1* locus of *Arabidopsis* disrupt the perception of phototropic stimuli. *Plant Cell* **7**, 473–485.

NPH3

Motchoulski, A.V. & Liscum, E. (1999) *Arabidopsis* NPH3: a NPH1 photoreceptor-interacting protein essential for phototropism. *Science* **286**, 961–964.

NPH4

Harper, R.M., Stowe-Evans, E.L., Luesse, D.R. *et al.* (2000) The *NPH4* locus encodes the auxin response factor ARF7, a conditional regulator of differential growth in aerial *Arabidopsis* tissue. *Plant Cell* **12**, 757–770.

Stowe-Evans, E.L., Harper, R.M., Motchoulski, A.V. & Liscum, E. (1998) NPH4, a conditional modulator of auxin-dependent differential growth responses in *Arabidopsis. Plant Physiology* **118**, 1265–1275.

Case study 7.5: Photoperiodic control of flowering

General review

Lin, C. (2000) Photoreceptors and the regulation of flowering time. *Plant Physiology* **123**, 39–50.

Signalling between leaves and the shoot apex

Bernier, G., Havelange, A., Houssa, C., Petitjean, A. & Lejeune, P. (1993) Physiological signals that induce flowering. *Plant Cell* **5**, 1147–1155.

Weller, J.L., Reid, J.B., Taylor, S.A. & Murfet, I.C. (1997) The genetic control of flowering in pea. *Trends in Plant Science* **2**, 412–418.

TOC1

Strayer, C., Oyama, T., Schultz, T.F. *et al.* (2000) Cloning of the *Arabidopsis* clock gene *TOC1*, an autoregulatory response regulator homolog. *Science* **289**, 768–771.

ELF3

McWatters, H.G., Bastow, R.M., Hall, A. & Millar, A.J. (2000) The *ELF3 zeitnehmer* regulates light signalling to the circadian clock. *Nature* **408**, 716–720.

Photoreceptors mediating the photoperiodic response

Guo, H., Yang, H., Mockler, T.C. & Lin, C. (1998) Regulation of flowering time by *Arabidopsis* photoreceptors. *Science* **279**, 1360–1363.

Mockler, T.C., Guo, H., Yang, H., Duong, H. & Lin, C. (1999) Antagonistic actions of *Arabidopsis* cryptochromes and phytochrome B in the regulation of floral induction. *Development* **126**, 2073–2082.

Suárez-López, P. & Coupland, G. (1998) Plants see the blue light. *Science* **279**, 1323–1324.

CONSTANS

Putterill, J., Robson, F., Lee, K., Simon, R. & Coupland, G. (1995) The *CONSTANS* gene of *Arabidopsis* promotes flowering and encodes a protein showing similarities to zinc finger transcription factors. *Cell* **80**, 847–857.

Samach, A., Onouchi, H., Gold, S.E. *et al.* (2000) Distinct roles of CONSTANS target genes in reproductive development of *Arabidopsis. Science* **288**, 1613–1616.

CHAPTER 8

Environmental information other than light

The previous chapter discussed the profound influence of light on plant development. However, light is only one of many environmental factors to regulate development in plants. Others include gravity, mechanical stimuli, temperature, nutrients, water, oxygen and atmospheric carbon dioxide (CO_2). Whilst some of these factors have a major impact on plant form, others play a more subtle role. For example, nutrient-rich patches in the soil promote lateral root formation; responses to gravity regulate the direction of growth in both shoots and roots; but the only direct developmental response to atmospheric CO_2 appears to be a change in stomatal density (the density is lower when CO_2 concentration is high). Similarly, some environmental factors induce responses in a wide range of species, while others only affect plants adapted to specialized environments. For example, about three-quarters of all species tested show morphogenic responses to mechanical stimuli, whereas only aquatic plants and those adapted to flood-prone environments display direct developmental responses to root anoxia.

This chapter considers four examples of developmental responses to diverse environmental stimuli, all of these occur in a wide range of species. Case study 8.1 considers plant responses to gravity. Case study 8.2 analyses the effects of mechanical stimuli on development. Case study 8.3 discusses how the development of the root system responds to uneven nutrient supply. Finally, case study 8.4 considers the promotion of flowering by **vernalization** (exposure to prolonged cold).

Case study 8.1: Gravitropism

The direction of gravity is an invariant indicator of orientation and often induces a bending response called **gravitropism**. In most angiosperms different parts of the shoot and root systems grow at different angles with respect to the gravity vector. Primary shoots and roots usually grow vertically, up and down, respectively. Often, secondary and tertiary shoots and roots initially orient closer to the horizontal and grow away from the main axis, subsequently turning to more vertical growth. Even more dramatic changes in growth with respect to the gravity vector are widely observed in plants with trailing growth habits. It is common for shoots of trailing plants initially to grow upwards, this is followed by a period of downwards growth, and subsequently by a resumption of upwards growth—producing an S-shaped shoot.

These responses are likely to have adaptive significance. For example, in germinating seedlings, vertical growth of the primary shoot and root increases the chance of successful establishment. Throughout the life cycle, the diverse orientations of second- and higher-order shoots and roots increase the volume of space occupied by the shoot and root systems, enhancing the absorption of light and the assimilation of water and nutrients.

The mechanism of gravitropism differs between herbaceous and woody organs (Fig. 8.1). During primary growth, gravitropic bending is caused by differential cell elongation. If a vertically growing shoot is placed horizontally, cell elongation on the upper side declines while cell elongation on the lower side increases. Hence the shoot bends upwards. The opposite pattern of changes causes a horizontally placed root to bend downwards.

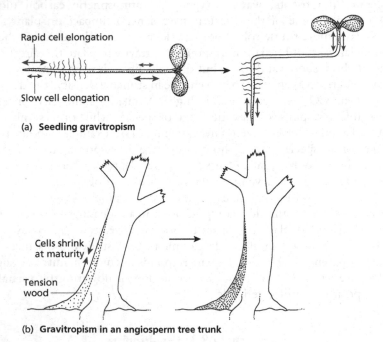

(a) **Seedling gravitropism**

(b) **Gravitropism in an angiosperm tree trunk**

Fig. 8.1 Gravitropism in seedlings and tree trunks. (a) In a horizontally placed seedling, cell elongation increases on the upper side of the root and decreases on the lower side, causing the root to bend downwards. The opposite pattern of changes in elongation rates causes the shoot to bend upwards. (b) When an angiosperm tree trunk is tilted away from the vertical, for example by a gust of wind, tension wood forms on the upper side of the trunk. Cells in tension wood shrink at maturity, pulling the trunk upright. ((b) adapted from Mattheck, 1990.)

Woody shoots orient relative to gravity by the formation of **reaction wood**. For example, if a strong gust of wind pushes an angiosperm tree away from the vertical, the tree produces **tension wood** on the upper side of the trunk. Cells in tension wood shrink at maturity and over several years the process pulls the top of the trunk upright. In similar circumstances, conifers produce **compression wood** on the lower side of the trunk. Compression wood expands as it matures, pushing the top of the trunk upwards. Reaction wood is also produced by branches and maintains the branches' orientation against the tendency to sag as their weight increases.

Most research on gravitropism has focused on the primary root and shoot of seedlings. This case study discusses the perception and transduction of the gravitational signal.

Perception of gravity

There is convincing evidence that gravity perception involves the sedimentation of starch-filled plastids called **amyloplasts**. Amyloplasts are sufficiently large and dense to fall through the cytoplasm after a change in organ orientation. Treatments that deplete the plant of starch reduce amyloplast sedimentation and impair gravitropism. For example, starch-deficient mutants of *Arabidopsis* have reduced gravitropic responses.

Amyloplast sedimentation occurs in specialized cells called **statocytes**. In roots, the statocytes are in the columella, i.e. the central root cap (Fig. 8.2). Columella cells are relatively unvacuolated and their endoplasmic reticulum lies around the cell periphery. The nucleus is usually at the end of the cell closest to the root meristem while amyloplasts lie in the central and distal regions. By comparison, lateral root cap cells are highly vacuolated; they have a central nucleus surrounded by amyloplasts; and the endoplasmic reticulum is distributed throughout the cytoplasm (Fig. 8.2). The function of the columella in sensing gravity has been confirmed by laser ablation experiments. In *Arabidopsis*, laser ablation of the columella virtually abolishes root gravitropism, whereas roots with an ablated lateral root cap remain gravitropic.

In the stem, statocytes form the innermost layer of the cortex, i.e. the layer immediately outside the vascular bundles. This tissue is called the **shoot endodermis** or sometimes the **starch sheath** (Fig. 8.3). The requirement of the shoot endodermis in gravitropism is demonstrated by *scarecrow* (*scr*) and *short-root* (*shr*) mutants of *Arabidopsis*. These lack a shoot endodermis and have agravitropic shoots. *scr* and *shr* were originally identified because they lack a distinct root endodermis (see case study 4.3). However, the mutants develop a columella and have normal root gravitropism.

Signal transduction in the statocyte

The immediate mechanism linking amyloplast movement to gravitropic curvature is unknown. One possibility is that amyloplasts induce localized endoplasmic reticulum responses as they settle into the endoplasmic reticulum

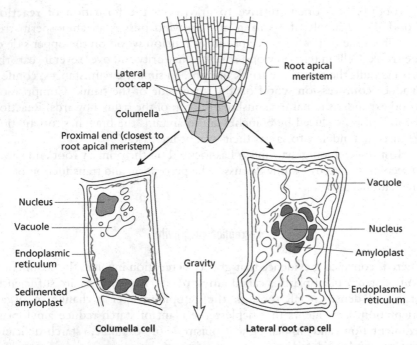

Fig. 8.2 The ultrastructure of columella and lateral root cap cells in *Arabidopsis*. Columella cells, unlike lateral root cap cells, have sedimentable amyloplasts (shown in red) and a peripheral endoplasmic reticulum. (Adapted from Sack, 1997.)

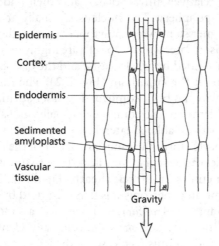

Fig. 8.3 Longitudinal section of an *Arabidopsis* hypocotyl. In the hypocotyl, graviperception appears to occur in the endodermis where there are sedimentable amyloplasts (shown in red). The structure of stems is more complex, but the position of the endodermis relative to the vascular tissue is retained. (Adapted from Fukaki *et al.*, 1998.)

that lines the periphery of statocytes (Fig. 8.2). Another possibility is suggested by the observation that actin filaments appear to connect amyloplasts to the plasma membrane. Therefore, amyloplast movement may be detected by deformation of the cytoskeleton or of the membrane.

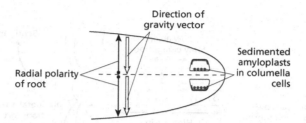

Fig. 8.4 The relationship between the gravity vector and radial polarity in a horizontal root. Columella cells in the upper and lower halves of the root have different responses to gravity (see text). This could be achieved if cells compared the gravity vector with the radial polarity of the root. For example if the outer (red) and inner (uncoloured) halves of the columella cell relative to the root radial axis responded differently to amyloplast sedimentation.

Gravitropism is associated with electrical events at statocyte plasma membranes. Immediately after a root is re-oriented relative to gravity, columella cells on the lower side depolarize, i.e. the electrical potential across their plasma membranes falls, whereas columella cells on the upper side hyperpolarize, i.e. membrane potentials rise. Amyloplast sedimentation may therefore initiate ion movements across the plasma membrane. Re-orienting the root also induces complex pH changes in columella cells.

The above responses to gravistimulation can only occur if cells can detect whether they are in the upper or lower half of the root. Cells may achieve this by comparing the direction of gravity with the polarity of the root's radial axis. For example, if a cell is in the upper half of the root, the gravity vector points towards the root core, whereas if a cell is in the lower half, the gravity vector points away from the core. This comparison could be made cell-autonomously if individual columella cells were polarized along the radial axis (Fig. 8.4).

Signalling between statocytes and other cells

In the root, gravity perception occurs in the columella whereas gravitropic bending takes place behind the root meristem in the zone of cell elongation. In the stem, gravity perception occurs in statocytes in the shoot endodermis but changes in cell elongation rates are clearest in the epidermis and peripheral layers of the cortex. Although controversial, it is likely that intercellular signalling between statocytes and the elongating tissues involves the redistribution of auxin (Fig. 8.5).

Direct measurements of endogenous auxin and the movement of externally applied, radiolabelled auxin indicate that tipping a vertically growing shoot to a horizontal position leads to an increase in auxin concentration in the lower side of the stem. There is a corresponding decrease in auxin in the upper side. Under normal circumstances, auxin promotes cell elongation in the shoot. Therefore, the auxin redistribution increases cell elongation in the lower half of the stem and decreases elongation in the upper half, causing the stem to bend upwards. This is very similar to the Cholodny–Went mechanism of phototropism (see case study 7.4).

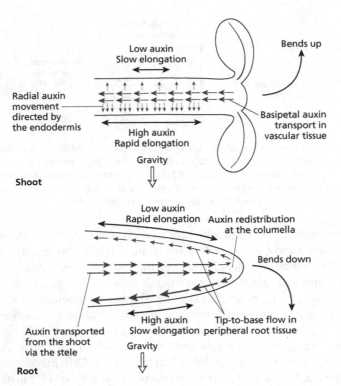

Fig. 8.5 Models for the redistribution of auxin during shoot and root gravitropism. Gravity-sensitive auxin redistribution occurs at the sites of gravity perception: the endodermis in the shoot and the columella in the root. This leads to a rise in auxin concentration in the lower halves of horizontally placed shoots and roots, and a fall in auxin concentration in the upper halves. The effects of the resulting auxin gradient are different in the two organs, causing shoots to bend up and roots to bend down. It is likely that the gravitropic response also involves changes in auxin sensitivity (see text).

Similar measurements show that moving a vertically growing root to the horizontal also increases auxin concentration in the lower side and decreases auxin concentration in the upper side. For example, expression of the auxin-inducible gene, *IAA2*, in *Arabidopsis* is stronger in the lower half of a horizontally placed root than in the upper half. Unlike in the shoot, under normal circumstances auxin inhibits cell elongation in the root. Auxin redistribution in the root therefore enhances cell elongation in the upper half but reduces elongation in the lower half, causing the root to bend downwards.

The mechanism by which gravity perception affects auxin distribution is probably different in shoots and roots. As described in Chapter 4, auxin is synthesized in the shoot tip and is transported basipetally in the polar auxin transport stream which occurs in cells associated with the vascular system. In addition, there is likely to be some radial auxin movement from the vascular tissue outwards to supply peripheral stem tissue with auxin. Since such a radial auxin flux would pass through statocytes in the shoot endodermis, the supply of auxin to peripheral tissues could be modulated by gravity perception (Fig. 8.5).

In the root, there are two auxin streams. Auxin arriving from the shoot flows towards the root tip through the stele. At the tip, auxin is redirected into more peripheral root tissues through which it flows back up the root (see Fig. 4.5). Given that gravity perception occurs in the root tip, it is probable that the flux of auxin through the tip is regulated to direct more auxin to the lower half of a horizontally placed root (Fig. 8.5).

Mutants in auxin signalling or transport confirm that auxin is pivotal to the gravitropic response. For example, *nonphototropic hypocotyl4* (*nph4*) mutants of *Arabidopsis* have reduced cellular response to auxin and reduced phototropism and gravitropism (see case study 7.4). The *auxin resistant1* (*aux1*) and *ethylene insensitive root1* (*eir1*) mutants of *Arabidopsis* relate gravitropism in the root to auxin transport (*eir1* is also called *pin2* and *agr1*). Auxin transport occurs by the influx and efflux of auxin through successive cells (see Fig. 4.5), with the direction of transport determined by the polarized location of efflux carriers in the plasma membrane. *AUX1* is thought to encode a subunit of the auxin influx carrier of root cells, while *EIR1* encodes a subunit of the efflux carrier of root cells. Loss of either gene results in agravitropic roots. Gravitropism is also blocked when auxin transport inhibitors are applied to wild-type roots.

The agravitropic phenotypes of *aux1* and *eir1* mutants demonstrate that auxin flux is required for root gravitropism but do not confirm that changes in the auxin flux are involved. Indeed, the *aux1* agravitropic phenotype can be rescued by the uniform application of the membrane permeable auxin, NAA (α-napthaleneacetic acid), which is able to enter cells in the absence of a functional influx carrier. This suggests that merely supplying auxin to the elongation zone is sufficient for gravitropism. However, it is possible that even in this situation an auxin gradient is created by differential activity of the efflux carrier. This is supported by experiments demonstrating that a change in root orientation changes the accumulation of the EIR1 protein across the root tip, through a decrease in EIR1 in peripheral cells in the upper side. The change in EIR1 accumulation should decrease the concentration of auxin flowing into peripheral tissue in the top half of the root, strongly supporting a role for auxin redistribution in gravitropism.

Auxin redistribution is unlikely to explain all aspects of gravitropism because in some circumstances it is not measurable. For example, as a horizontally placed root bends back towards the vertical, the root tip often overshoots and bending reverses. The process repeats leading to progressively damped oscillations around the vertical. There is no evidence that the oscillations are matched by oscillating auxin concentrations across the root. This has led to the suggestion that changes in auxin sensitivity are also involved in the gravitropic response.

Effects of light on gravitropism

Gravitropism can be modulated by light. For example, the formation of an apical hook in etiolated seedlings represents a complex gravitropic response. As the seedling grows, cells move basipetally through the hook and the hook itself moves progressively closer to the shoot apex. Regions of stem more apical than the hook orient downwards, those more basal than the hook orient upwards, and those passing through the hook change from one pattern of gravitropism to the other. When an etiolated seedling is exposed to light, the overall pattern of gravitropism changes so that the stem straightens. Mutational analysis in *Arabidopsis* suggests that both crypto-chromes and phytochromes mediate hook opening (see case study 7.2).

Case study 8.2: Thigmomorphogenesis

Plants receive a variety of mechanical stimuli, including disturbance by animals, wind and rain, and pressure from obstacles impeding growth. Col-lectively, developmental responses to mechanical stimuli are called **thigmo-morphogenesis**. Mechanical stimuli can also induce a bending response called **thigmotropism**. For example, tendrils, petioles and/or stems of climbing plants bend towards a touch stimulus and so wrap around supports; in contrast roots of many species avoid obstructions by bending away from touch stimuli.

As is the case for gravitropism, the mechanism of thigmomorphogenesis differs between herbaceous and woody organs. During primary growth, mechanical stimuli reduce longitudinal cell elongation and enhance radial cell expansion, leading to shorter and stockier plants. Hence seedlings grow-ing through compacted soil develop shorter, broader shoots and roots than seedlings growing through loose soil. Compacted soil also leads to tightening of the shoot apical hook. Together, these changes are likely to increase the ability of the seedling to withstand longitudinal pressure as it pushes through the soil (Fig. 8.6a).

Above the soil, the effects of wind exposure and other mechanical stimuli on cell expansion result in shorter and stockier stems. In addition, mechan-ical stimuli also lead to a decrease in the growth rate of the shoot, an increase in lignification and a reduction in leaf size (Fig. 8.6b). These changes strengthen the shoot and reduce the area presented to the wind. Mechanical stimulation of the shoot also alters the development of the root system. Roots in regions of the soil parallel to the prevailing wind, i.e. windward or leeward of the shoot, are typically thicker and more numerous than those in regions perpendicular to the wind. These changes probably improve the ability of the root system to anchor the shoot.

Mechanical stimuli affect the form of woody shoots and roots by influ-encing the production of new wood. For example, tree trunks thicken most quickly along the axis parallel to the prevailing wind, strengthening the trunk against gusts from that direction. Unlike the production of specialized reac-tion wood during gravitropism, mechanical stimuli usually lead to an increase in the synthesis of normal wood.

Early responses to mechanical stimuli

Thigmomorphogenesis can be induced by extremely mild touch stimuli. For example, spraying greenhouse-grown tomato plants with water for 10

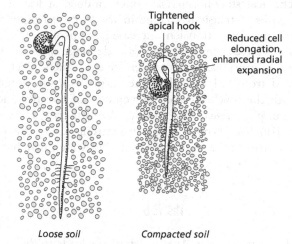

(a)   **Thigmomorphogenesis in seedlings beneath the soil**

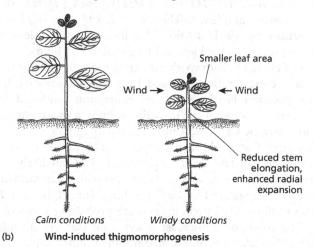

(b)   **Wind-induced thigmomorphogenesis**

Fig. 8.6 Thigmomorphogenesis during primary development. (a) Seedlings germinated in compacted soil are shorter and broader than seedlings germinated in loose soil. This is due to a reduction in the rate of elongation and an increase in the rate of radial expansion of cells in the radicle and hypocotyl/epicotyl. Seedlings in compacted soil also have a tightened apical hook. (b) Wind exposure likewise reduces cell elongation and increases cell radial expansion, resulting in shorter, stockier plants. In addition, windy conditions result in a reduction in leaf size.

seconds a day reduces their eventual height to about 60% of that of control plants. Of course, under natural circumstances, plants are always exposed to moderate shaking by the wind and plants growing outdoors are correspondingly shorter and stockier than those in greenhouses.

The mechanism by which plants perceive mechanical stimuli is unclear but it is likely to involve responses to deformation of the plasma membrane or cytoskeleton. It is worth noting that gravity perception also requires plants to respond to mechanical stimuli: in this case, to the minute forces generated by the movement of amyloplasts (see case study 8.1).

Mild mechanical stimuli, such as touch, induce at least two immediate cellular responses: a transient change in the electrical potential across the plasma membrane, and a transient increase in cytosolic Ca^{2+} concentration. These responses have been detected in a wide range of tissue types, suggesting that many or most plant cells are sensitive to touch. Electrical events can be transmitted from cell to cell in the form of action potentials and are associated with the touch-induced movements of sensitive and carnivorous plants, for example *Mimosa* spp. and the Venus-flytrap, *Dionaea muscipula*, respectively. Touch-induced changes in cytosolic Ca^{2+} concentration promote the expression of several genes, some of which may function in thigmomorphogenesis (see below).

The *TOUCH* genes

Briefly touching or spraying *Arabidopsis* plants leads to the rapid accumulation of mRNA from five *TOUCH* (*TCH*) genes (*TCH1–5*). Within 10 minutes of a touch stimulus, mRNAs encoded by the *TCH* genes increase in concentration by 10–100-fold. The increase is transient and mRNA concentrations return to background levels after 1–3 hours.

Two lines of evidence from work with cultured cells indicate that the rise in cytosolic Ca^{2+} concentration induced by a mechanical stimulus promotes *TCH* gene expression. Firstly, *TCH* gene expression is induced by an increase in the Ca^{2+} concentration in the growth medium, a treatment that leads to a rapid rise in cytosolic Ca^{2+}. Secondly, chelating agents, which diminish the supply of external Ca^{2+}, inhibit the expression of at least some *TCH* genes.

The proteins encoded by *TCH1*, *TCH2* and *TCH3* are likely to be involved in transducing the Ca^{2+} signal. The TCH1 protein is a **calmodulin**, a signalling protein that is regulated by Ca^{2+} binding. The *TCH2* and *TCH3* genes encode calmodulin-related proteins and are probably also involved in Ca^{2+} signalling. The ability of an increase in cytosolic Ca^{2+} to stimulate the production of Ca^{2+}-receptor proteins may amplify cellular responses to touch.

TCH4 encodes a xyloglucan endotransglycosylase (XET), an enzyme involved in cell wall synthesis and/or modification. XETs can cut and ligate xyloglucan strands, which crosslink cellulose microfibrils in the cell wall, and in doing so regulate the physical properties of the wall. Since the direction and extent of cell expansion depends substantially on the structure of the cell wall, TCH4 may mediate thigmomorphogenic changes in cell expansion patterns.

Detailed tissue-level expression patterns for *TCH3* and *TCH4*, determined using promoter–reporter gene fusions and by immunolocalization of the TCH3 and TCH4 proteins, show that both genes are widely expressed during plant development. For example, *TCH3* is expressed strongly at the junctions between leaves and the stem, and both *TCH3* and *TCH4* are highly expressed in regions of cell expansion. These sites are likely to be subject to high mechanical stress or strain.

It is likely that the *TCH* genes function in other processes, such as responses to heat shock, cold shock and darkness. Like mechanical stimuli, temperature shocks cause an immediate increase in cytosolic Ca^{2+} concentration, and both temperature shocks and darkness strongly induce *TCH* gene expression.

Ethylene

It is probable that ethylene is involved in some aspects of thigmomorphogenic signal transduction. Ethylene synthesis increases in response to mechanical stimuli such as rubbing or pressure, and exposure to ethylene induces thigmomorphogenic responses. For example, etiolated *Arabidopsis* seedlings that are exposed to ethylene display the **triple response**, characterized by a tightened apical hook, reduced stem elongation and increased stem thickness. As described above, the same responses are induced in seedlings by the pressure of compacted soil (Fig. 8.6a). Compacted soil increases ethylene production due to mechanical stimulation, and also decreases the rate at which ethylene diffuses away from the plant. Similarly, green shoots exposed to ethylene develop a shorter and stockier morphology.

Cells perceive ethylene through a small family of plasma membrane receptors. In *Arabidopsis*, these are encoded by five genes (in abbreviated form): *ETR1*, *ETR2*, *ERS1*, *ERS2* and *EIN4*. Dominant mutations in any one of these genes cause reduced ethylene binding by the receptor encoded by that gene, and the loss of ethylene responses by the plant. The loss of ethylene responses is surprising since such mutants still have four genes encoding wild-type receptors. That the ethylene receptor genes are functionally redundant is demonstrated by the fact that null mutations in any one of them have no phenotypic effects. However, strikingly, plants carrying null mutations in multiple members of the family show a constitutive triple response in the absence of ethylene.

Taken together, these data suggest a counterintuitive model for ethylene perception in which the receptors are active when ethylene is absent and are inactivated by ethylene binding (Fig. 8.7). According to this hypothesis, reduced ethylene binding by a receptor carrying a dominant mutation prevents the inactivation of receptor signalling by ethylene, resulting in a constitutive 'no ethylene' signal and the loss of ethylene responses. In contrast, in multiple null mutants, receptor activity is significantly reduced even in the absence of ethylene, resulting in constitutive ethylene responses.

The ethylene receptors regulate the activity of the CONSTITUTIVE TRIPLE RESPONSE1 (CTR1) protein, which was defined by loss-of-function mutations that cause a constitutive triple response. This suggests that the CTR1 protein acts to inhibit ethylene responses, and that in the absence of ethylene, the ethylene receptors activate CTR1 signalling (Fig. 8.7). Ethylene binding prevents receptor activity and therefore CTR1 signalling, allowing ethylene responses to occur. CTR1 is homologous to mammalian RAF kinases, which initiate kinase signalling cascades that end with the regulation of transcription factors. Therefore, it is likely that ethylene downregulates a kinase signalling cascade initiated by CTR1. Downstream effects of this putative signalling cascade include inhibition of the EIN3 family of transcription factors, which promote the expression of ethylene-inducible genes (Fig. 8.7).

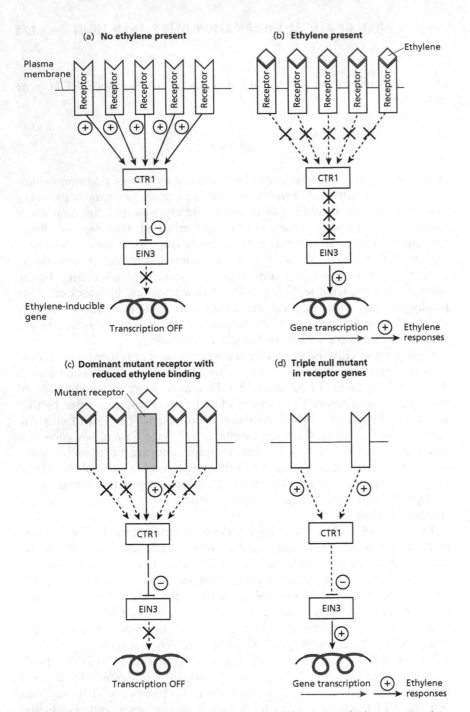

Fig. 8.7 Ethylene signal transduction. (a) In the absence of ethylene, multiple receptors activate CTR1. This induces a putative signalling cascade leading to the inhibition of the EIN3 family of transcription factors, preventing the transcription of ethylene-inducible genes. (b) Ethylene binding inactivates receptor signalling, allowing the EIN3 family to induce the transcription of ethylene-inducible genes. (c) Dominant mutations that reduce receptor–ethylene binding cause constitutive activation of the CTR1 pathway, preventing ethylene responses. (d) Null mutations in multiple receptor genes reduce the number of active receptors such that ethylene responses are no longer inhibited in the absence of ethylene. Arrows accompanied by a 'plus' sign indicate positive regulation. Lines ending in a bar and accompanied by a 'minus' sign indicate negative regulation.

The relationship between ethylene and the *TCH* genes is unclear. *TCH* gene expression is not induced by supplying plants with ethylene, nor is it altered by ethylene insensitivity caused by dominant mutations in the *ETR1* ethylene receptor gene. These results suggest that *TCH* genes either function upstream of ethylene in a touch-induced signalling pathway or that they act in a separate pathway.

Case study 8.3: Effects of uneven nutrient supply on root development

Local enrichment of the soil by decaying organic matter and local impoverishment by root absorption mean that in natural environments soil nutrients are distributed unevenly. In many species, root development responds to uneven nutrient supply by increased root growth in nutrient-rich patches and decreased root growth in other regions. The major nutrients required by plants are nitrogen, phosphorus and potassium. For many species, uneven distributions of nitrogen in the form of nitrate (NO_3^-) or ammonium (NH_4^+), or of phosphorus in the form of phosphate (PO_4^{3-}) have dramatic effects on root architecture. In contrast, responses to an uneven distribution of potassium ions are relatively rare.

Responses to nutrient distribution also vary between species. Of those species tested, about one-third show no developmental response to uneven nutrient supply, while the magnitude of the response varies greatly in the remainder. In general, fast-growing annuals display the most dramatic responses. This may reflect the fact that a temporary nutrient-rich patch has more significance to a short-lived plant.

Nitrate supply and *Arabidopsis* root architecture

Lateral roots of *Arabidopsis* display two contrasting responses to high nitrate (Fig. 8.8). Firstly, a high overall supply of nitrate reduces lateral root extension throughout the root system. This effect appears to be a response to overall nitrate sufficiency because lateral root extension in nitrate-sufficient conditions is suppressed even in parts of the root system that are not growing in high nitrate. The second response to high nitrate is observed when the overall nitrate supply is low. Under these conditions, lateral root growth increases 2–3-fold in local nitrate-rich patches. This is accompanied by a small decrease in lateral root growth everywhere else.

Root growth requires nutrients absorbed from the soil and photosynthates transported from the shoot. It has been proposed that uneven nutrient distributions may affect root architecture indirectly by local metabolic effects. For example, if a lateral root from a nitrate-starved plant grew into a nitrate-rich region, the abundant nitrate would enable that root to grow more rapidly. In doing so, the lateral root would increase its own consumption of photosynthates and reduce the photosynthates available to other lateral roots, inhibiting their growth. This is known as a **metabolic mechanism**.

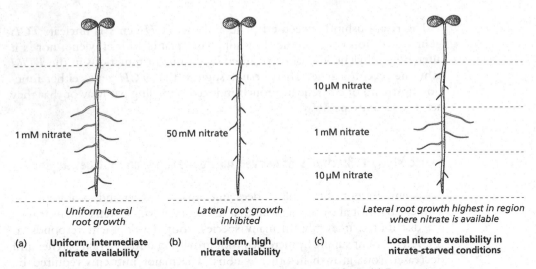

Fig. 8.8 The effect of nitrate availability on lateral root growth in *Arabidopsis*. (a) Root systems exposed to a uniform, intermediate nitrate concentration display uniform lateral root growth. (b) Lateral root growth is inhibited by uniformly high nitrate availability. (c) In nitrate-starved conditions, lateral root growth is promoted in regions where nitrate is available. (Adapted from Zhang & Forde, 2000.)

It is now clear that a metabolic mechanism is not required to control the response of the *Arabidopsis* root system to nitrate. *Arabidopsis* mutants severely deficient in nitrate reductase are effectively unable to use nitrate as a source of nitrogen. Therefore, nitrate cannot alter the nitrogen status of mutant roots. Even so, both of the responses to nitrate described above still occur. A similar observation has been made using nitrate-reductase-deficient lines of tobacco, which retain an overall inhibition of root growth imposed by high nitrate supply.

The alternative to a metabolic mechanism is that nitrate acts as a signalling molecule, regulating root growth by affecting developmental signalling pathways. The *ARABIDOPSIS NITRATE REGULATED1* (*ANR1*) gene, discussed below, is a likely component of such a signalling pathway.

ANR1

The *ANR1* gene was identified in a screen for mRNAs that increase in concentration in response to nitrate. *ANR1* mRNA is undetectable in the roots of nitrate-starved plants but accumulates within half an hour in roots supplied with nitrate. *ANR1* encodes a MADS domain transcription factor.

Transgenic plants with reduced *ANR1* levels do not respond to locally high nitrate supply with increased lateral root growth. However, such plants still display the overall inhibition of lateral root growth imposed by globally high nitrate. Indeed, inhibition of lateral root growth in *ANR1*-deficient plants is hypersensitive to nitrate. Wild-type plants show inhibition of lateral root growth when supplied with 10 mM of nitrate, whereas inhibition occurs at 0.1 mM nitrate in *ANR1*-deficient plants.

These results can be explained by assuming that lateral root growth is

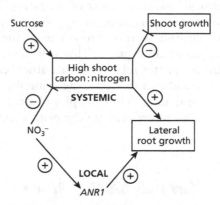

Fig. 8.9 The local and systemic effects of nitrate on lateral root growth. Lateral root growth is promoted systemically by a high carbon : nitrogen ratio in the shoot (top). This ratio is reduced by nitrate, consequently a rise in nitrate concentration inhibits lateral root growth. Nitrate promotes local transcription of the *ANR1* gene which functions to promote the growth of individual lateral roots (bottom). Arrows accompanied by a 'plus' sign indicate a positive effect. Lines ending in a bar and accompanied by a 'minus' sign indicate a negative effect. (Adapted from Zhang *et al.*, 1999.)

controlled by two antagonistic mechanisms (Fig. 8.9). In this model, local nitrate enhances the growth of individual lateral roots by an *ANR1*-dependent mechanism. However, global nitrate causes an overall inhibition of lateral root growth by a mechanism independent of *ANR1*. The growth of individual lateral roots will therefore depend on the relative contributions of the local and global mechanisms. In *ANR1*–deficient plants, local promotion of lateral root growth does not occur. Hence the global inhibition of root growth by nitrate is detectable at a lower nitrate concentration.

Adaptive significance of responses to nitrate

The global inhibitory effect of nitrate on lateral root growth is likely to be part of the mechanism that regulates the allocation of resources between shoots and roots. In many plants, the relative growth rates of the root and shoot systems appear to be regulated by the internal carbon : nitrogen ratio of the shoot. The concentration of carbon, in the form of sucrose, is an indicator of photosynthetic capacity. Likewise, the concentration of nitrogen, for example as nitrate, is an indicator of nutrient availability. Often, conditions that result in a high carbon : nitrogen ratio promote root over shoot development, whereas conditions that result in a low carbon : nitrogen ratio promote shoot over root development (Fig. 8.9). In this way the ability to photosynthesize is balanced with the ability to assimilate nutrients. Hence when overall nitrate supply is low, more resources are directed towards root growth and the absorption of nitrate, whereas in nitrate-sufficient conditions, root growth is reduced and more resources are available for shoot growth and hence photosynthesis.

The adaptive significance of altering root development in response to nitrate-rich patches has been debated. Nitrate is a very mobile ion and so a single root will eventually assimilate most of the nitrate in a relatively large volume of soil. However, this perspective ignores competition for nitrate between plants. An increase in root growth in a nitrate-rich patch will increase the rate of nitrate absorption. This will increase the amount of nitrate that the plant assimilates relative to that absorbed by neighbouring plants. Furthermore, in addition to nitrate, decaying organic matter releases slower moving ions such as ammonium and phosphate. Root proliferation in response to nitrate may therefore increase the plant's ability to compete for other nutrients.

Case study 8.4: Vernalization

As discussed in case study 7.5, seasonal patterns of development can be regulated by changes in the photoperiod. Often, however, changes in temperature are also important in relating development to the time of year. For example, in deciduous trees, low autumn temperatures can stimulate the development of over-wintering buds. Typically, such buds will not resume growth unless exposed to a further and prolonged period of cold: a process called **vernalization**. This mechanism prevents occasional warm days from inducing premature growth in winter. In a similar way, vernalization can be required for subsequent germination and/or flowering, ensuring that these processes occur in spring or summer.

This case study considers the promotion of flowering by vernalization, discussing vernalization requirements, and the perception and transduction of the vernalization signal.

Vernalization requirements

The flowering response to vernalization varies between and within species. Most biennial species have an **absolute** vernalization requirement and will not flower unless exposed to near-freezing temperatures for days or weeks. Other plants show a **facultative** vernalization response. In such cases, vernalization reduces flowering time, but unvernalized plants do eventually flower. In such plants, the reduction in flowering time is often proportional to the length of the cold treatment. A facultative vernalization response is typical of **winter annuals**, such as winter varieties of wheat. These varieties are sown in the autumn and are vernalized during the winter as imbibed seeds or as seedlings. Spring-sown varieties of wheat flower rapidly without vernalization and show little or no vernalization response.

In some perennials, the vernalization requirement is reset each year. For example, the Michaelmas daisy (*Aster novi-belgii*) requires vernalization before it will flower. After flowering is complete, new shoots grow from the base of the plant and these shoots require fresh vernalization to induce them to flower. This ensures that the new shoots will not flower until the following year.

Perception of cold

The perception of cold and subsequent cellular responses have been studied in the context of freezing tolerance. In a process known as **acclimation**, plants exposed to several days of low but non-freezing temperatures become less susceptible to damage in freezing conditions. The initial fall in temperature induces an immediate, transient increase in cytosolic calcium ion concentration. Subsequently, the activity of a number of genes is induced, several of which, for example the *TCH* genes, are also induced by other environmental stimuli (see case study 8.2). Eventually, acclimation depends on physiological changes such as the accumulation of cryoprotectants (sugars and polyamines) in the cytoplasm and changes to the composition of cell membranes.

A fall in temperature makes cell membranes more rigid and several experiments suggest that the perception of cold may depend on changes in membrane fluidity. For example, at 25°C, chemical treatment of alfalfa cell cultures to increase membrane rigidity leads to an immediate Ca^{2+} influx, expression of the cold acclimation-specific gene *CAS30*, and subsequent freezing tolerance. In contrast, at 4°C, chemical treatments that lower membrane rigidity prevent these phenomena.

The phenotype of the *high response to osmotic stress1* (*hos1*) mutant of *Arabidopsis* suggests that acclimation and vernalization are induced by overlapping mechanisms. Acclimation-related genes are induced more rapidly and to a greater extent by cold exposure in *hos1* mutants than in wild-type plants, suggesting that the *HOS1* gene negatively regulates the response to cold. Furthermore, *hos1* mutants express a cold-inducible reporter gene at temperatures up to 19°C, compared to 7°C for wild type, indicating that the *HOS1* gene acts to repress cold responses at temperatures above 7°C. Interestingly, *hos1* plants flower substantially earlier than wild type and vernalization accelerates flowering to a greater extent in wild-type plants than in mutants. This suggests that constitutive cold-responsive signalling in *hos1* mutants leads to a constitutive vernalized status.

HOS1 is expressed throughout the plant and encodes a ubiquitin–protein ligase similar to that encoded by *COP1* (see case study 7.2). These enzymes act by attaching ubiquitin to other proteins, targeting them for destruction. Therefore, *HOS1* may repress cold-responsive signalling by accelerating the degradation of proteins that promote cold responses. As described in case study 7.2, the COP1 protein shuttles between the cytoplasm and the nucleus in a light-dependent manner: the protein is cytoplasmic in the light and nuclear in the dark. The HOS1 protein also shuttles between the cytoplasm and the nucleus, in this case in response to temperature. HOS1 is cytoplasmic at normal growth temperatures but accumulates in the nucleus at low temperatures.

Site of cold perception in vernalization

The site at which cold is perceived in the flowering response to vernalization has been investigated by selectively chilling different parts of field pennycress

(*Thlaspi arvense*), a winter annual with an absolute vernalization require-ment. When using intact plants, the only effective flower-inducing treatment is to chill the shoot apex, suggesting that this is the normal site of cold perception in the vernalization response (Fig. 8.10). However, further ex-periments show that most or all parts of the plant respond to vernaliza-tion. For example, if a leaf cutting is taken from a vernalized plant, shoots regenerated from the cutting will flower (Fig. 8.10d). Likewise, experiments with other species show that vernalization of isolated roots, and even of cells growing in tissue culture, can accelerate flowering in regenerated plants.

These data suggest that vernalization acts at a local level. Vernalizing a single leaf, for example, promotes flowering in shoots regenerated from the leaf but does not induce flowering in unvernalized sections of the original plant. Consistent with this, grafting experiments indicate that vernalization does not result in the production of a flower-promoting signal. In most cases, grafting an unvernalized shoot onto a vernalized plant does not induce the grafted shoot to flower.

Maintenance of the vernalized state

There is often a delay of weeks or months between the end of a cold period and the onset of flowering. During this time, the plant maintains its vernal-ized state despite extensive growth in warm conditions. This 'memory' of vernalization and the local nature of the vernalization response can be explained if daughter cells inherit the vernalized or unvernalized status of the mother cell.

One candidate for encoding the vernalization status of the cell is the pattern of DNA methylation. The DNA of both plants and animals can be methylated on the cytosine residue of CG dinucleotides. In plants, cytosine methylation also occurs in C(any nucleotide)G triplets. DNA methylation in the promoter region of genes usually inhibits gene expression and therefore provides a mechanism for gene regulation. Furthermore, the pattern of DNA methylation is normally preserved through DNA replication and is heritable between cell generations, i.e. it is lineage-dependent. Consistent with the need for vernalization by each generation, methylation patterns are usually reset by sexual reproduction.

Several observations support a role for DNA methylation in vernalization. Firstly, prolonged exposure to low temperatures leads to a substantial drop in the proportion of methylated DNA. For example, keeping *Arabidopsis* seed-lings at 8°C for 4 weeks reduces DNA methylation to around 86% of the level measured in unvernalized seedlings. Secondly, treating plants with 5-azacyti-dine (5-azaC), which results in DNA demethylation, induces early flowering in varieties that respond to vernalization but does not affect flowering time in varieties with no vernalization response. For example, 5-azaC promotes flowering in winter wheat but not in spring wheat. Similarly, vernalization-sensitive *Arabidopsis* lines can be engineered to show reduced methylation by antisense expression of the gene encoding the methyltransferase enzyme.

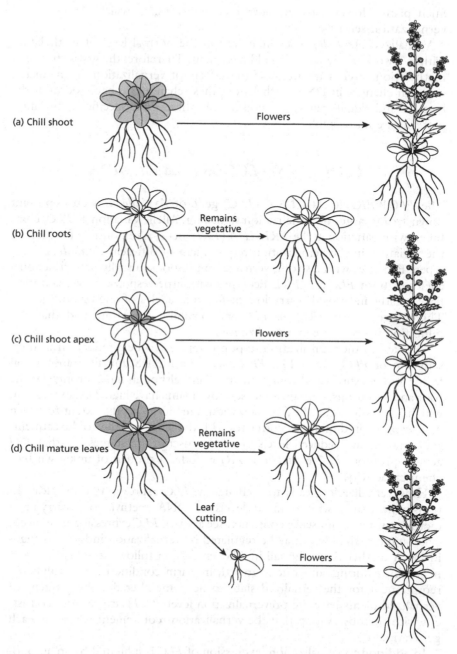

Fig. 8.10 The effects of vernalizing different regions of *Thlaspi arvense* (field pennycress). *Thlaspi arvense* has an absolute vernalization requirement and will not flower unless subject to a cold period. Flowering can be induced by specifically chilling the whole shoot (a) or only the shoot apex (c), but not by chilling only the roots (b) or only mature leaves (d). However, a leaf cutting taken from a chilled leaf will flower (d). These results suggest that vernalization acts at a local level in the plant.

Such plants flower early in short-day conditions, in which *Arabidopsis* is vernalization-sensitive.

Vernalized *Arabidopsis* seedlings regain the normal level of methylation within 1 week of the end of a cold treatment. Therefore, the gross change in methylation level is transient. If the effects of vernalization are mediated through changes in DNA methylation, these changes probably occur at the level of individual genes and presumably these gene-specific methylation patterns are more stable.

FLOWERING LOCUS C and *FRIGIDA*

The *FLOWERING LOCUS C (FLC)* gene of *Arabidopsis* is an important downstream target of vernalization signalling. Allelic variation at *FLC*, combined with variation at the *FRIGIDA (FRI)* locus, is responsible for most of the variation in the vernalization requirement of different *Arabidopsis* ecotypes. Ecotypes with a strong vernalization response usually have functional alleles of both *FLC* and *FRI*. Ecotypes with little response to vernalization flower early and usually carry loss-of-function alleles of one or both genes. This suggests that *FRI* and *FLC* act to inhibit flowering, and that this inhibition is overcome by vernalization.

The key to the vernalization response appears to be a reduction in transcription of *FLC* (Fig. 8.11). *FLC* encodes a MADS domain transcription factor and is expressed throughout the plant, although most strongly at the root and shoot tips. Vernalization-sensitive plants have high *FLC* expression, which is greatly reduced by vernalization. Furthermore, the extent to which *FLC* expression diminishes correlates with the length of the cold treatment. *fri* mutants have reduced levels of *FLC* transcription, suggesting that *FRI* acts to promote *FLC* expression. *FRI* encodes a protein of unknown biochemical activity.

How vernalization leads to a change in *FLC* expression is not clear. In transgenic plants with reduced levels of DNA methylation (carrying a methyltransferase antisense construct, see above), *FLC* expression is reduced. This suggests that *FLC* may be regulated by vernalization-induced changes in DNA methylation. The fall in *FLC* expression following vernalization is maintained during subsequent growth in warm conditions. This fulfils the requirement for the vernalized state to be 'remembered'. The progeny of vernalized plants have the prevernalization level of *FLC* expression, consistent with the observation that the vernalization requirement is reset in each generation.

In addition to vernalization, expression of *FLC* is inhibited by an autonomous flower-promoting pathway (Fig. 8.11). This is discussed in detail in case study 9.3, which considers the integration of the various mechanisms that control flowering.

Downstream targets of *FLC* include the *AGAMOUS-LIKE20* gene (*AGL20*, also called *SOC1*) and probably the *FLOWERING LOCUS T (FT)* gene (Fig. 8.11). Both of these genes function to promote flowering, and both are also regulated by the CONSTANS protein during the photoperiodic

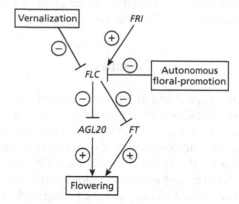

Fig. 8.11 Vernalization, *FLC* and flowering. Expression of the *FLC* gene inhibits flowering through the inhibition of flower-promoting genes such as *AGL20* and *FT*. FLC expression is promoted by the action of the *FRI* gene, and is inhibited by vernalization and by an autonomous floral-promotion pathway. The regulation of flowering is discussed further in Chapter 9 (see Fig. 9.11). Arrows accompanied by a 'plus' sign indicate positive regulation. Lines ending in a bar and accompanied by a 'minus' sign indicate negative regulation.

response (see case study 7.5). *AGL20* transcription is initially low in plants with functional *FLC* alleles and rises after vernalization, as *FLC* expression falls. This suggests that *FLC* inhibits *AGL20* transcription, perhaps through direct action of the FLC transcription factor. By reducing *FLC* expression, therefore, vernalization leads to an increase in *AGL20* transcription, and so promotes flowering. Regulation of *FT* expression by *FLC* has not been demonstrated directly. However, *FT* expression is promoted by the autonomous flower-promoting pathway and it is thought that this may be because *FT* expression is inhibited by *FLC*.

Evolution of vernalization-insensitive *Arabidopsis* ecotypes

Arabidopsis is widely distributed throughout the northern hemisphere and ecotypes collected from different locations vary greatly in their requirement for vernalization. In general, there is a strong relationship between a southern habitat and lack of a vernalization requirement. This is not surprising since a need for vernalization would be a major selective disadvantage in locations with mild winters.

As described above, lack of a vernalization requirement is normally associated with loss-of-function alleles of *FRI* and/or *FLC*. A survey of 40 ecotypes has shown that the most common cause of vernalization independence is loss of *FRI* function. Most vernalization-independent ecotypes carry one or other of two partial deletion alleles of *FRI*. These ecotypes, therefore, represent two independent groups that both arose from ancestors carrying complete, functional *FRI* genes and which were probably vernalization-sensitive. This suggests a northern origin for the species. Interestingly, vernalization-independent ecotypes within the two groups are very widely distributed. For example, the same 376 base pair deletion in *FRI* is found in ecotypes collected in France and in Japan. Since *Arabidopsis* is an agricultural weed, this may reflect dispersal by humans.

Conclusions

As discussed in this chapter and in Chapter 7, development is regulated by diverse environmental factors in ways that enhance the survival and/or reproductive ability of the plant. Usually, environmental information affects the orientation, extent and/or timing of organ development. For example, the direction of shoot and root growth is regulated by the directions of both gravity and light; the extent of stem elongation is regulated by the intensities of both light and mechanical stimuli; and the timing of flowering is regulated by both the photoperiod and vernalization.

Environmental responses depend on a complex interaction between environmental and internal information. Obviously, responses to environmental cues depend on the modification of internal information. However, it is also clear that internal information modulates the response to environmental factors. In many plants, for example, the primary shoot grows upwards while lateral shoots initially grow closer to the horizontal. The gravitropic response of an individual shoot therefore depends on internal information about the position of the shoot on the plant.

The response to one aspect of the environment is often also affected by other classes of environmental information. The modulation of gravitropism by light provides a good example. The gravitropic response of etiolated seedling shoots is changed by illumination, resulting in the opening of the apical hook. Similarly, lateral shoots of plants exhibiting shade escape usually grow closer to the vertical than lateral shoots of plants growing in full sunlight.

The next chapter discusses the integration of internal and environmental information necessary for coordinated development.

Further reading

Case study 8.1: Gravitropism

Reviews

Chen, R., Rosen, E. & Masson, P.H. (1999) Gravitropism in higher plants. *Plant Physiology* 120, 343–350.

Dolan, L. (1998) Pointing roots in the right direction: the role of auxin transport in response to gravity. *Genes and Development* 12, 2091–2095.

Evans, M.L. (1991) Gravitropism: interaction of sensitivity modulation and effector redistribution. *Plant Physiology* 95, 1–5.

Rosen, E., Chen, R. & Masson, P.H. (1999) Root gravitropism: a complex response to a simple stimulus? *Trends in Plant Science* 4, 407–412.

Sack, F.D. (1991) Plant gravity sensing. *International Review of Cytology* 127, 193–252.

Tasaka, M., Takehide, K. & Fukaki, H. (1999) The endodermis and shoot gravitropism. *Trends in Plant Science* 4, 103–107.

Gravity perception

Sack, F.D. (1997) Plastids and gravitropic sensing. *Planta* **203**, S63–S68.

Effects of scr and shr on shoot gravitropism

Fukaki, H., Wysocka-Diller, J., Kato, T. *et al.* (1998) Genetic evidence that the endodermis is essential for shoot gravitropism in *Arabidopsis thaliana*. *Plant Journal* **14**, 425–430.

Tasaka, M., Takehide, K. & Fukaki, H. (1999) The endodermis and shoot gravitropism. *Trends in Plant Science* **4**, 103–107.

AUX1

Bennett, M.J., Marchant, A., Green, H.G. *et al.* (1996) *Arabidopsis AUX1* gene: a permease-like regulator of root gravitropism. *Science* **273**, 948–950.

Marchant, A., Kargul, J., May, S.T. *et al.* (1999) *AUX1* regulates root gravitropism in *Arabidopsis* by facilitating auxin uptake within root apical tissues. *EMBO Journal* **18**, 2066–2073.

EIR1 (PIN2, AGR1)

Chen, R., Hilson, P., Sedbrook, J. *et al.* (1998) The *Arabidopsis thaliana AGRAVITROPIC 1* gene encodes a component of the polar-auxin-transport efflux carrier. *Proceedings of the National Academy of Science, USA* **95**, 15112–15117.

Luschnig, C., Gaxiola, R.A., Grisafi, P. & Fink, G.R. (1998) EIR1, a root-specific protein involved in auxin transport, is required for gravitropism in *Arabidopsis thaliana*. *Genes and Development* **12**, 2175–2187.

Müller, A., Guan, C., Galweiler, L. *et al.* (1998) AtPIN2 defines a locus of *Arabidopsis* for root gravitropism control. *EMBO Journal* **17**, 6903–6911.

Utsuno, K., Shikanai, T., Yamada, Y. & Hashimoto, T. (1998) *AGR*, an *Agravitropic* locus of *Arabidopsis thaliana*, encodes a novel membrane-protein family member. *Plant Cell Physiology* **39**, 1111–1118.

Interaction between light signals and gravitropism

Hangarter, R.P. (1997) Gravity, light and plant form. *Plant, Cell and Environment* **20**, 796–800.

Case study 8.2: Thigmomorphogenesis

General

Jaffe, M.J. (1973) Thigmomorphogenesis: the response of plant growth and development to mechanical stimulation. *Planta* **114**, 143–157.

Mechanical stimuli and cytosolic Ca^{2+} concentration

Haley, A., Russell, A.J., Wood, N. *et al.* (1995) Effects of mechanical signaling on plant cell cytosolic calcium. *Proceedings of the National Academy of Science, USA* **92**, 4124–4128.

The TCH genes

Antosiewicz, D.M., Polisensky, D.H. & Braam, J. (1995) Cellular localization of the Ca^{2+} binding TCH3 protein of *Arabidopsis*. *Plant Journal* **8**, 623–636.

Antosiewicz, D.M., Purugganan, M.M., Polisensky, D.H. & Braam, J. (1997) Cellular localization of *Arabidopsis* xyloglucan endotransglycosylase-related proteins during development and after wind stimulation. *Plant Physiology* **115**, 1319–1328.

Braam, J. & Davis, R.W. (1990) Rain-, wind-, and touch-induced expression of calmodulin and calmodulin-related genes in *Arabidopsis*. *Cell* **60**, 357–364.

Braam, J., Sistrunk, M.L., Polisensky, D.H. *et al.* (1997) Plant responses to environmental stress: regulation and functions of the *Arabidopsis TCH* genes. *Planta* **203**, S35–S41.

Ethylene signalling

Bleecker, A.B. (1999) Ethylene perception and signaling: an evolutionary perspective. *Trends in Plant Science* **4**, 269–274.

Case study 8.3: Effects of nutrient supply on root development

Reviews

Leyser, O. & Fitter, A. (1998) Roots are branching out in patches. *Trends in Plant Science* **3**, 203–204.

Robinson, D. (1994) The responses of plants to non-uniform supplies of nutrients. *New Phytologist* **127**, 635–674.

Zhang, H. & Forde, B.G. (2000) Regulation of *Arabidopsis* root development by nitrate availability. *Journal of Experimental Botany* **51**, 51–59.

ANR1

Zhang, H. & Forde, B.G. (1998) An *Arabidopsis* MADS box gene that controls nutrient-induced changes in root architecture. *Science* **279**, 407–409.

Further analysis of the Arabidopsis response to nitrate

Zhang, H., Jennings, A., Barlow, P.W. & Forde, B.G. (1999) Dual pathways for regulation of root branching by nitrate. *Proceedings of the National Academy of Science, USA* **96**, 6529–6534.

Nitrate and tobacco root development

Scheible, W.R., Lauerer, M., Schulze, E.D., Caboche, M. & Stitt, M. (1997) Accumulation of nitrate in the shoot acts as a signal to regulate shoot–root allocation in tobacco. *Plant Journal* **11**, 671–691.

Case study 8.4: Vernalization

Review

Michaels, S.D. & Amasino, R.M. (2000) Memories of winter: vernalization and the competence to flower. *Plant, Cell and Environment* **23**, 1145–1153.

Membrane fluidity

Örvar, B.L., Sangwan, V., Omann, F. & Dhindsa, R.S. (2000) Early steps in cold sensing by plant cells: the role of actin cytoskeleton and membrane fluidity. *Plant Journal* **23**, 785–794.

HOS1

Ishitani, M., Xiong, L., Lee, H., Stevenson, B. & Zhu, J-K. (1998) *HOS1*, a genetic locus involved in cold-responsive gene expression in *Arabidopsis*. *Plant Cell* 10, 1151–1161.

Lee, H., Xiong, L., Gong, Z. *et al.* (2001) The *Arabidopsis HOS1* gene negatively regulates cold signal transduction and encodes a RING finger protein that displays cold-regulated nucleo-cytoplasmic partitioning. *Genes and Development* 15, 912–925.

Localized chilling experiments

Metzger, J.D. (1988) Localization of the site of perception of thermoinductive temperatures in *Thlaspi arvense* L. *Plant Physiology* 88, 424–428.

DNA methylation

Burn, J.E., Bagnall, D.J., Metzger, J.D., Dennis, E.S. & Peacock, W.J. (1993) DNA methylation, vernalization, and the initiation of flowering. *Proceedings of the National Academy of Science, USA* 90, 287–291.

Finnegan, E.J., Genger, R.K., Kovac, K., Peacock, W.J. & Dennis, E.S. (1998) DNA methylation and the promotion of flowering by vernalization. *Proceedings of the National Academy of Science, USA* 95, 5824–5829.

FLC and FRI

Johanson, U., West, J., Lister, C. *et al.* (2000) Molecular analysis of *FRIGIDA*, a major determinant of natural variation in *Arabidopsis* flowering time. *Science* 290, 344–346.

Lee, H., Suh, S-S., Park, E. *et al.* (2000) The AGAMOUS-LIKE 20 MADS domain protein integrates floral inductive pathways in *Arabidopsis*. *Genes and Development* 14, 2366–2376.

Sheldon, C.C., Rouse, D.T., Finnegan, E.J., Peacock, W.J. & Dennis, E.S. (2000) The molecular basis of vernalization: the central role of FLOWERING LOCUS C (FLC). *Proceedings of the National Academy of Science, USA* 97, 3753–3758.

CHAPTER 9

The coordination of development

Plant development is continuous and modular, a combination that allows enormous plasticity of morphology. Continuous development is achieved by means of apical meristems which act by producing a series of modules (Fig. 9.1). In the root, a module can be considered to be the apical meristem and the root behind it extending to the first lateral branch. In the shoot, a module can be considered to be a leaf, an axillary meristem and an internode, which, collectively, are called a **phytomer**. Successive phytomers can vary in morphology; for example, phytomers produced at different stages of shoot development can have different sorts of leaf. Furthermore, axillary vegetative meristems have the potential to grow out into lateral shoots, reiterating the development of the primary shoot axis, and axillary floral meristems develop into flowers (Chapter 1). The flower is a determinate shoot in which internodes are extremely compact, leaves are replaced by floral organs and axillary meristems do not develop.

Plasticity is achieved through the integration of information from a variety of sources to regulate when and how many of each module is produced. This chapter discusses how the shoot apical meristem (SAM) achieves continuous development and how information is integrated to regulate module production by the meristem. It considers information relating to the developmental and physiological state of the plant, and to the environment.

The initiation of the SAM and the maintenance SAM size and structure are discussed in case study 9.1. Case studies 9.2 and 9.3 then consider the regulation of the activity of the SAM and of the sequence of modules it produces. Case study 9.2 analyses the transition from embryonic to post-embryonic development which occurs at germination. Case study 9.3 considers the sequence of modules produced during post-germination development, analysing the transitions from juvenile to adult and from adult to reproductive development. The morphology of the shoot system also depends on the extent of development at vegetative axillary SAMs. This is considered in case study 9.4, which discusses the control of shoot branching.

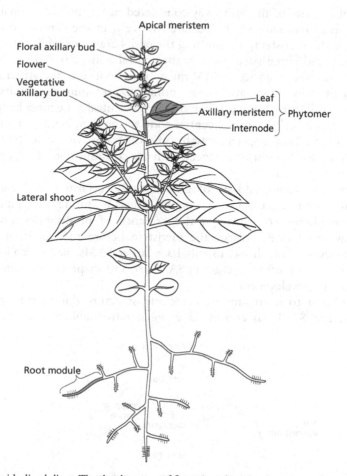

Apical meristem

Floral axillary bud

Flower

Vegetative
axillary bud

Lateral shoot

Leaf

Axillary meristem

Internode

Phytomer

Root module

Fig. 9.1 An idealized dicot. The development of flowering plants is reiterative and modular. The shoot apical meristem produces a series of phytomers (pink), each consisting of a leaf, axillary meristem and internode. Axillary meristems may develop into vegetative axillary buds, which can reiterate the development of the primary shoot, or into floral axillary buds, which develop into flowers. In the root, a module can be considered to be the root apical meristem along with the root behind it extending to the first lateral branch (pink).

Case study 9.1: Initiation and maintenance of the shoot apical meristem

As described above, continuous development of the shoot depends on the apical meristem. The initiation of the shoot apical meristem (SAM) during embryogenesis was described in Chapter 1 and discussed in more detail in Chapter 4 (case study 4.2). Chapter 1 also described the longitudinal and radial structure of the SAM (see Fig. 1.6). Longitudinally, the SAM is divided into an outer, layered region called the **tunica** that overlies an inner region called the **corpus**. Radially, the SAM is divided into a **central zone** of slowly dividing relatively large cells, and a **peripheral zone** of rapidly dividing smaller cells from which leaf primordia are initiated. Beneath the central zone there is a region called the **pith meristem** that initiates the stem pith.

Clonal analysis of the SAM was considered in Chapter 3. Clonal sectors derived from chimeric SAMs indicate that cells in the central zone act as initials for the meristem, replenishing the peripheral zone whilst maintaining a pool of 'undifferentiated' cells in the central zone. To allow continuous shoot development, an active SAM must maintain its size and structure as its constituent cells divide and organ primordia are initiated on its flanks. Therefore, cell growth and proliferation in the central zone must be balanced with the exit of cells from the central zone to the peripheral zone or to the pith meristem. Similarly, cell growth and proliferation in the peripheral zone and pith meristem must be balanced with the consumption of cells by organ initiation (Fig. 9.2a).

Direct observation and histological analysis confirm that the size and structure of an active SAM is roughly constant during the initiation of successive phytomers, although the presence of presumptive primordial cells may temporarily enlarge the meristem before the initiation of each leaf. However, adult shoots usually have larger SAMs than juvenile shoots, and there are also often changes in SAM size and shape on the transition to reproductive development.

In addition to maintaining its size and structure during normal development, the SAM has considerable regenerative ability (Fig. 9.2b). For

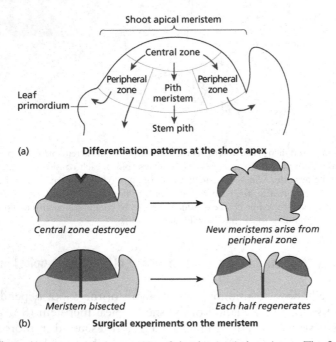

(a) **Differentiation patterns at the shoot apex**

Central zone destroyed   →   *New meristems arise from peripheral zone*

Meristem bisected   →   *Each half regenerates*

(b) **Surgical experiments on the meristem**

Fig. 9.2 The maintenance and regeneration of the shoot apical meristem. The figure shows longitudinal cross-sections through the shoot apex. (a) Differentiation patterns at the shoot apex. Cells leaving the central zone of the shoot apical meristem adopt peripheral zone or pith meristem identity, according to their position. Cells in the peripheral zone give rise to leaves and peripheral stem tissues, and cells in the pith meristem give rise to the stem pith. (b) Surgical experiments on the meristem. If the central zone of the meristem is destroyed, new meristems regenerate from the peripheral zone. If the meristem is bisected, each half regenerates into a complete meristem.

example, if the central zone of the SAM is destroyed, some cells in the peripheral zone adopt central zone fates and one or more functional SAMs regenerate. Similarly, if a SAM is bisected it will regenerate into two meristems of normal size.

Genetic regulation of SAM initiation and maintenance

Arabidopsis mutants with disrupted SAM activity have been used to identify genes required for normal SAM function and to characterize their wild-type roles. Prominent among these are *SHOOTMERISTEMLESS (STM)*, *WUSCHEL (WUS)*, *CLAVATA1* and *CLAVATA3 (CLV1* and *CLV3)*, which are discussed below.

SHOOTMERISTEMLESS

Complete loss-of-function *stm* seedlings have no SAM and have fused cotyledons (Fig. 9.3a), indicating that *STM* is required for SAM formation in the embryo. Eventually, *stm* seedlings produce a few leaves in a disorganized pattern (Fig. 9.3b), suggesting that the mutants form short-lived adventitious SAMs and, therefore, that *STM* is not absolutely required for meristem initiation.

Mutants carrying weak, partial loss-of-function *stm* alleles produce shoots but these eventually terminate as the cells of the SAM become vacuolated and cease to divide. Often, there is an ectopic leaf primordium or a group of fused primordia in the centre of the terminated meristem. Similarly, mutant floral meristems frequently terminate in a cluster of fused floral organs before a full flower has developed. As discussed in Chapter 1, floral meristems represent modified vegetative SAMs. Overall, therefore, the phenotype of weak *stm* mutants suggests that low *STM* activity compromises the maintenance of the SAM.

STM is a *KNOX* gene, i.e. it encodes a homeodomain transcription factor related to the maize KNOTTED1 protein. It is expressed in the embryonic shoot apex from the late globular stage onwards and its expression eventually becomes restricted to the developing SAM. Post-embryonically, *STM* is expressed in all vegetative SAMs, in inflorescence meristems and in floral meristems. As with other *KNOX* genes, *STM* expression disappears from the founder cells of each leaf primordium shortly before the primordium becomes visible (Fig. 9.4). *STM* expression is similarly absent in floral organ primordia. Consequently, consistent with the determinate nature of floral development, *STM* expression in the floral meristem ends as the carpels are initiated (except for some localized *STM* transcription at the sites of ovule development). Two other *KNOX* genes, called *KNAT1* and *KNAT2*, have been described in *Arabidopsis* and, like *STM*, they are active in the SAM but not in leaf primordia.

The maintenance of SAM development by *STM* appears to be mediated by the inhibition of the expression of a gene called *ASYMMETRIC LEAVES1* *(AS1)*. As described in case study 5.3, the inhibition of *KNOX* expression in

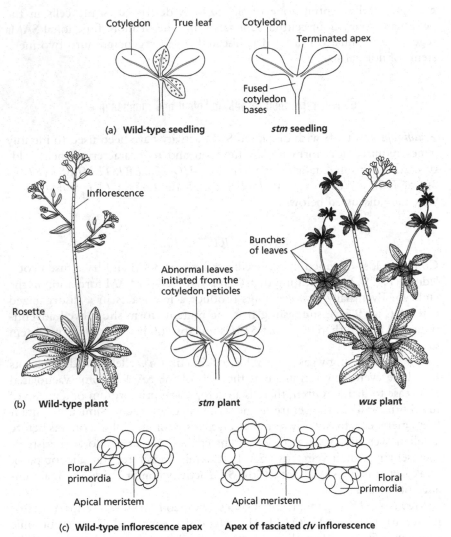

(a) **Wild-type seedling**   *stm* **seedling**

(b)   **Wild-type plant**   *stm* **plant**   *wus* **plant**

(c) **Wild-type inflorescence apex**   **Apex of fasciated** *clv* **inflorescence**

Fig. 9.3 *Arabidopsis* mutants with disrupted shoot apical meristem development. (a) A wild-type seedling and a seedling homozygous for a strong *stm* allele. The *stm* mutant lacks a shoot apical meristem and has partially fused cotyledons. (b) A wild-type plant, a strong *stm* mutant showing leaves that have developed from the hypocotyl area, and a *wus* plant. The eventual production of leaves by the *stm* mutant suggests the formation of short-lived, adventitious shoot apical meristems from or near the cotyledon petioles. The *wus* phenotype arises because of the failure to maintain the shoot apical meristem, followed by development at adventitious and/or axillary meristems: giving a stop–start pattern of growth. (c) Wild-type and *clv* inflorescence apices, showing the apical meristems and youngest floral primordia. The apical meristem of *clv* mutants enlarges during shoot development, resulting in a fasciated (ribbon-like) stem. (Adapted in part from Barton & Poethig, 1993; Clark *et al.*, 1995; Laux *et al.*, 1996.)

developing leaves requires homologues of the *PHANTASTICA* (*PHAN*)-gene of *Antirrhinum*. These encode MYB-like transcription factors and are expressed in a complementary pattern to *KNOX* genes, i.e. at the sites of leaf initiation and in leaf primordia but not in other parts of the SAM. In

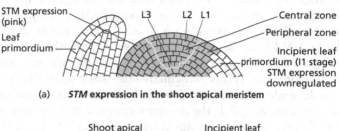

(a) **STM expression in the shoot apical meristem**

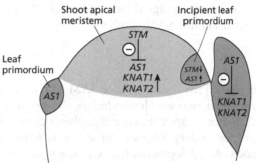

(b) **The interactions between AS1 and the KNOX genes**

Fig. 9.4 The role of *STM* in maintaining the apical meristem. (a) *STM* expression in the shoot apical meristem. *STM* is expressed throughout the meristem but its expression is downregulated at the sites of leaf initiation. *STM* is not expressed in the leaf primordia. (b) The interactions between *AS1* and the *KNOX* genes. *STM* inhibits the expression of *AS1*, and *AS1* inhibits the expression of *KNAT1* and *KNAT2*. *STM* expression in the meristem (pink), results in the inhibition of *AS1* expression, thereby allowing the expression of *KNAT1* and *KNAT2*. Down-regulation of *STM* expression at the sites of leaf initiation, allows the upregulation of *AS1* expression. In the leaf primordium, *AS1* is expressed and inhibits the expression of *KNAT1* and *KNAT2*. Lines ending in a bar and accompanied by a 'minus' sign indicate negative regulation.

Arabidopsis, *AS1* has been identified as a *PHAN* homologue. *as1* mutants produce deeply lobed leaves on which shoots sometimes develop, and mutant leaf primordia ectopically express *KNAT1* and *KNAT2* but do not express *STM*. This suggests that *AS1* specifically inhibits *KNAT1* and *KNAT2* expression. In contrast, whereas wild-type embryos express *AS1* only in the cotyledon primordia, embryos of strong *stm* mutants also express *AS1* at the presumptive SAM site. This suggests that *STM* inhibits *AS1* expression.

The significance of ectopic *AS1* expression in *stm* mutants is dramatically demonstrated by the almost complete suppression of the *stm* phenotype in the *stm as1* double mutant. Whereas *stm* mutants germinate without a SAM, *stm as1* double mutants germinate with a functional SAM and have a vegetative phenotype almost identical to that of *as1* single mutants. These data suggest a model in which *STM* maintains SAM activity by repressing *AS1* (Fig. 9.4b). In *stm* loss-of-function mutants, *AS1* is expressed ectopically at the SAM. The ectopic *AS1* represses *KNAT1* and *KNAT2* expression, resulting in the complete loss of *KNOX* activity and hence the failure of SAM development. In *stm as1* double mutants, however, *KNAT1* and *KNAT2* expression continues at the SAM, allowing SAM development and shoot production. In this model, wild-type leaf initiation proceeds by the following sequence of events: *STM* is downregulated at the sites of leaf initiation; this allows the upregulation of *AS1*, which acts to downregulate *KNAT1* and *KNAT2* in the presumptive leaves (Fig. 9.4b).

The antagonistic relationship between *STM* and *AS1* acts to separate SAM from leaf development. This is particularly interesting given the evidence for the mutual dependence of SAM activity and the formation of leaf primordia. As discussed in case study 5.1, the development of adaxial tissues in the leaf is thought to require a polarizing signal from the SAM. In turn, leaves may produce a signal that maintains SAM activity.

WUSCHEL

Like *STM*, *WUS* is needed for indeterminate SAM activity. Loss-of-function *wus* mutants initiate about two true leaves before the seedling SAM terminates in an enlarged, flattened apex. Eventually, adventitious SAMs initiate at the shoot apex, and/or axillary SAMs form at the base of the cotyledons. After variable periods of development, the secondary shoots also terminate and tertiary SAMs arise, reiterating the start–stop developmental pattern (Fig. 9.3b). Similarly, mutant flowers produce approximately normal sepal and petal whorls but usually terminate early in a single central stamen. These phenotypes indicate that *WUS* is required for meristem maintenance.

WUS encodes a homeodomain transcription factor but is not a member of the *KNOX* family. *WUS* expression is first detected at the 16-cell stage of embryogenesis, i.e. in the very early globular embryo, in a subset of cells in the apical half of the embryo. During the initiation of the SAM, *WUS* expression is progressively restricted to a cluster of cells in the corpus of the central zone (Fig. 9.5). It has been suggested that these cells produce a signal that specifies central zone identity in the overlying tissue. According to this theory, in *wus* mutants, the signal is absent and cells in the central zone adopt other identities, resulting in meristem termination. Interestingly, there is evidence that the quiescent centre of the root apical meristem plays a similar role in maintaining meristem activity (see box 9.1). The function and regulation of *WUS* is revealed further by experiments on *CLV1* and *CLV3*, discussed below.

CLAVATA1 and *CLAVATA3*

CLV1 and *CLV3* are required to restrict the size of the central zone. *clv1* mutants, *clv3* mutants and *clv1 clv3* double mutants have the same phenotype. Mutants germinate with an enlarged SAM and the SAM continues to enlarge throughout shoot development. In extreme cases, mutant meristems can reach about 1000 times the volume of the wild-type meristem. Because the *clv* SAM enlarges preferentially in one horizontal plane, adopting a Mars Bar shape (Fig. 9.3c), the stem below it becomes ribbon-like, a phenomenon called **fasciation**.

The floral meristems of *clv* mutants are also enlarged, and produce more floral organs per whorl than wild type. The most dramatically affected whorl

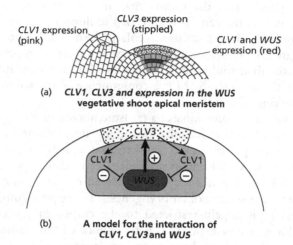

(a) **CLV1, CLV3 and expression in the WUS vegetative shoot apical meristem**

(b) **A model for the interaction of CLV1, CLV3 and WUS**

Fig. 9.5 The *WUS/CLV3/CLV1* loop. (a) *CLV1*, *CLV3* and *WUS* expression in the vegetative shoot apical meristem. *CLV3* is expressed in the outer two layers (the tunica) of the central zone (stippled red). *CLV1* is expressed in deeper layers (the corpus) of the central zone (pink). Both *CLV1* and *WUS* are expressed by a small group of cells in the corpus (red). (b) A model for the interaction of *CLV1*, *CLV3* and *WUS*. The model proposes that cells expressing the *WUS* gene produce a signal that promotes central zone identity and the expression of *CLV3* in the tunica of the shoot apical meristem. The CLV3 protein acts as a signal to deeper layers of the meristem (arrows), binding to the CLV1 protein and resulting in the inhibition of *WUS* expression. This establishes a feedback loop that maintains the size of the central zone. Arrows accompanied by a 'plus' sign indicate positive regulation. Lines ending in a bar and accompanied by a 'minus' sign indicate negative regulation. ((a) adapted from Bowman & Eshed, 2000; (b) adapted from Schoof *et al.*, 2000.)

Box 9.1: Maintenance of initial cell populations in the shoot and root apical meristems

There is an interesting parallel between the small group of *WUSCHEL* (*WUS*)-expressing cells in the shoot apical meristem (SAM) and the quiescent centre of the root apical meristem (RAM). As discussed in case study 9.1, *WUS*-expressing cells in the SAM prevent cells in the central zone from differentiating, thus maintaining the pool of initial cells necessary for indeterminate development. Laser ablation experiments on the *Arabidopsis* root suggest that the quiescent centre serves a similar function in the RAM.

The *Arabidopsis* RAM has a simple and invariant structure: a quiescent centre of four, non-dividing cells is surrounded by initial cells from which the main body of the root and the root cap are derived (see Fig. 1.5). Ablation of the whole quiescent centre leads to the displacement of dead quiescent centre cells towards the root tip and subsequent regeneration of the RAM (see case study 3.1). If just one or two cells from the quiescent centre are ablated, however, replacement of the dead cells is much slower and the effects of the ablation on the neighbouring initial cells can be observed. Under these circumstances, initial cells in contact with an ablated quiescent centre cell lose their initial cell characteristics and differentiate. This suggests that the quiescent centre produces a signal that maintains the pool of initial cells in the RAM, just as *WUS*-expressing cells appear to maintain the central zone of the SAM.

is the carpel whorl where the mean organ number is more than doubled, suggesting that it is the central region of the floral meristem that is most enlarged. Similarly, close analysis of enlarged *clv* SAMs shows that they consist of a greatly enlarged central zone surrounded by a peripheral zone of much greater than wild-type circumference but approximately wild-type width. Hence the enlarged SAM is a consequence of the failure to regulate the size of the central zone.

CLV1 encodes a receptor kinase: a transmembrane protein that responds to the binding of an extracellular ligand by intracellular kinase activity. Hence, the CLV1 protein relays cell-extrinsic signals to the intracellular signalling network. *CLV3* encodes a small, extracellular protein that is a ligand of CLV1. Both genes are first expressed in the developing SAM at the early heart-shaped stage of embryogenesis. In the post-embryonic SAM, *CLV1* expression is largely restricted to the corpus of the central zone, whereas *CLV3* is mainly expressed in the tunica of the central zone (Fig. 9.5). Therefore, CLV3 is likely to be an apoplastic signal from the tunica of the central zone to *CLV1*-expressing cells in the central zone corpus. Further evidence that CLV3 acts as a mobile signal comes from the analysis of *CLV3–clv3–clv3* chimeras, in which only the L1 cells, i.e. those in the outermost tunica layer, have a functional *CLV3* gene. These chimeric plants are phenotypically wild type, indicating that CLV3 produced in the L1 can restrict the size of the central zone in deeper layers of the meristem.

Interactions between *WUS*, the *CLV* genes and *STM*

The enlargement of the central zone in *clv1* and *clv3* mutants is accompanied by a greatly enlarged zone of *WUS* transcription. The expansion of the domain of *WUS* expression appears to be required for the development of enlarged SAMs in the *clv* mutants because the *clv1 wus* and *clv3 wus* double mutants have a *wus*-small, determinate meristem phenotype. These data suggest that a primary role for the *CLV* genes is to limit the domain of *WUS* expression. The observation that in the wild-type meristem, the region of *CLV1* expression encompasses the region of *WUS* expression (Fig. 9.5) is consistent with this idea.

To test the affect of *WUS* activity on the expression of the *CLV* genes, transgenic plants were generated in which *WUS* expression was activated in developing leaf primordia. This results in the cessation of leaf formation and also the ectopic expression of *CLV3*. This is consistent with the hypothesis discussed above in which *WUS* expression leads to the production of a signal that confers central zone identity, and suggests that one consequence of the signal is the induction of *CLV3* expression.

WUS, *CLV3* and *CLV1* therefore form elements of a feedback loop that maintains the size of the central zone (Fig. 9.5b). *WUS* expression in the corpus of the central zone induces *CLV3* expression in the overlying tunica. The CLV3 protein that is produced as a consequence activates CLV1 signalling in the corpus. CLV1 activation in turn limits the region of *WUS* expression.

What, then, is the relationship between the *WUS/CLV3/CLV1* loop and *STM*? As described above, *WUS* is first expressed in apical cells in the very early globular embryo; *STM* is first expressed in the apical region of the late globular embryo; and *CLV1* and *CLV3* are first expressed in the developing SAM at the early heart-shaped stage of embryogenesis. Interestingly, although *stm* embryos do not form a histologically distinguishable SAM, *CLV1* expression continues in the presumptive SAM region of *stm* mutants until the torpedo stage and *WUS* expression continues until late embryogenesis. Similarly, *STM* expression initiates as normal and continues in the appropriate position until late in the embryogenesis of *wus* mutants. These data suggest that the initiation of the *WUS/CLV3/CLV1* loop and *STM* expression occur independently in the embryo. In contrast, *WUS*, *CLV3* and *CLV1* are not expressed in the terminated apex of *stm* seedlings. Similarly, *STM* is not expressed in the terminated apex of *wus* seedlings, and *STM* expression spreads to encompass the enlarged SAM of *clv* mutants. This indicates at least some cross-talk between the *WUS/CLV3/CLV1* loop and *STM* expression during post-embryonic development. However, whereas the *WUS/CLV3/CLV1* loop is primarily concerned with maintaining a population of central zone cells, *STM* is not only required for SAM maintenance but is intimately involved in the regulation of leaf specification.

The *stm*, *wus* and *clv* mutants are far from alone in having disrupted SAM development. Other notable examples include the *mgoun1* (*mgo1*) and *mgoun2* (*mgo2*) mutants of *Arabidopsis*, which have enlarged SAMs and produce fewer leaves and other lateral organs than wild type. Unlike the enlarged SAMs of *clv* mutants, SAMs of *mgo* mutants have an approximately normal central zone but an enlarged peripheral zone. This suggests that the *MGO* genes are required for cells to make the transition from peripheral zone identity to leaf identity.

The SAM as a self-perpetuating structure

A critical feature of the mechanisms that maintain SAM activity is that they are self-perpetuating. For example, once established, the *WUS/CLV3/CLV1* loop will run independently, providing continual regulation of central zone size. Similarly, once leaf initiation has begun, the feedback loop between leaf formation and the maintenance of SAM activity by signals from leaf primordia is self-maintaining.

The mechanisms that maintain the SAM also show considerable redundancy. For example, in the absence of both *STM* and *AS1* function, SAM development appears to be maintained by the *KNAT1* and *KNAT2 KNOX* genes. Similarly, the development of short-lived shoots in *wus* mutants could be interpreted as indicating that other genes act redundantly with *WUS* to promote central zone identities. Consistent with this, *CLV3* is still expressed at a low level in *wus* mutants. Continuity of SAM activity is therefore achieved by a belts-and-braces approach in which redundant, self-referential mechanisms maintain SAM size and structure as successive modules in the shoot are initiated.

Despite the apparent autonomy of the SAM maintenance system, factors external to the meristem must impact significantly upon it. Dramatic changes in the activity, size and shape of the SAM occur throughout the life cycle of the plant. The relationships between these external factors and the SAM maintenance system are not currently well defined at the molecular level but they are central to achieving environmentally sensitive developmental plasticity.

Case study 9.2: Transition from embryonic to post-embryonic development

Shoot development can be divided into separate phases, which are characterized by the production of different modules by the shoot apical meristem (SAM). In the embryonic phase, the SAM, cotyledons and hypocotyl form, and, in many species, embryonic leaves are initiated by the SAM. In the post-embryonic phase, the seedling germinates and the SAM produces juvenile, adult and then reproductive modules. This case study and the next discuss the timing of the transitions between the phases of shoot development. This case study considers the regulation of the transition from embryonic to juvenile development, while case study 9.3 discusses the regulation of the transitions from juvenile to adult, and from adult to reproductive development.

Seed maturation and germination

In the vast majority of angiosperms there is a pause between seed formation and germination. Towards the end of seed development, the embryo and/or endosperm accumulate storage products—proteins, lipids and starch in proportions and amounts that depend on the species—after which the seed desiccates. This is called **seed maturation**.

There is convincing evidence that seed maturation is induced by abscisic acid (ABA). ABA concentration in the seed rises at the end of embryogenesis, reaches a maximum during the accumulation of storage products, and declines during desiccation. Treating developing seeds with ABA promotes the accumulation of storage products and the aquisition of desiccation tolerance. In contrast, mutant embryos deficient in ABA biosynthesis or ABA response often display reduced synthesis of storage products; desiccation intolerance; and **vivipary**, i.e. germination while the seed is still on the parent plant.

The control of germination varies extensively between species. It can depend on environmental factors, such as light, temperature and water availability; and on factors relating to the state of the seed or embryo, such as the time elapsed since the seed imbibed water. In many species, these factors appear to be mediated by gibberellins. For example, *gibberellic acid1* (*ga1*) mutants of *Arabidopsis* are unable to synthesize gibberellins and will not germinate unless treated with gibberellins. Furthermore, the induction of germination in *Arabidopsis* by light correlates with an increase in both gibberellin synthesis and sensitivity (see case study 7.1).

At germination, the pattern of shoot development changes dramatically as the shoot enters the juvenile phase. Study of the *LEAFY COTYLEDON1* (*LEC1*) and *PICKLE* (*PKL*) genes of *Arabidopsis* suggests that in addition to promoting germination, gibberellins inactivate embryonic developmental programmes, allowing progression to the juvenile phase.

LEAFY COTYLEDON1

Embryos of the *lec1* mutant of *Arabidopsis* accumulate lower than wild-type levels of seed storage proteins and cannot tolerate desiccation. This suggests that the *LEC1* gene is required for normal seed maturation. Furthermore, if *lec1* embryos are excised before desiccation and are grown in culture, the seedlings that develop have trichomes on the cotyledons (Fig. 9.6b). In wild-type *Arabidopsis*, leaves develop trichomes but cotyledons are hairless, suggesting that *LEC1* is also required to promote cotyledon over leaf identity.

LEC1 encodes a transcription factor homologous to eukaryotic CCAAT-box

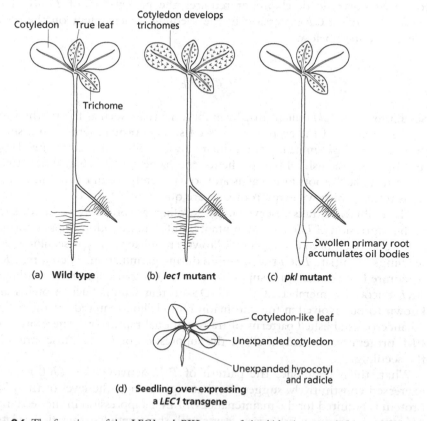

Fig. 9.6 The functions of the *LEC1* and *PKL* genes of *Arabidopsis* as revealed by mutants and transgene expression. (a) In wild-type plants, trichomes form on true leaves but not cotyledons. (b) In *lec1* mutants, cotyledons develop leaf characteristics such as trichomes. In contrast, seedlings overexpressing a *LEC1* transgene have embryonic characteristics such as cotyledon-like leaves (d). These data suggest that *LEC1* promotes embryonic development. (c) Seedling *pkl* mutants have embryonic characteristics such as the accumulation of oil bodies. This suggests that *PKL* is required to suppress embryonic traits in the seedling. ((b) adapted from Meinke, 1992; (c) adapted from Ogas *et al.*, 1997; (d) adapted from Lotan *et al.*, 1998.)

binding factors, which function as part of transcriptional activator complexes, and is expressed only during embryogenesis, in both the embryo and the endosperm. It is probable, therefore, that the LEC1 transcription factor activates the transcription of embryo-specific genes to promote seed maturation and cotyledon over leaf identity. Consistent with the temporal pattern of *LEC1* expression, *lec1* mutants have a wild-type phenotype after the seedling stage.

The function of *LEC1* has been investigated further by producing transgenic plants in which *LEC1* is expressed during both embryonic and post-embryonic development. Such plants have embryonic characteristics after germination (Fig. 9.6d). Seedlings are smaller than wild type and express seed storage protein genes, their cotyledons do not expand in the light, and they produce cotyledon-like leaves from which ectopic embryos sometimes develop. Therefore, *LEC1* transcription is not only necessary for normal embryogenesis, but also sufficient to induce embryonic traits during post-embryonic development. This suggests that the transition from embryonic to post-embryonic development requires the inactivation of *LEC1*. The prevention of *LEC1* expression in the seedling is dependent on the *PKL* gene, described below.

PICKLE

Seedlings of the *pkl* mutant display embryonic traits such as the synthesis of storage lipids and the expression of seed storage protein genes. In a small proportion of *pkl* mutants, the primary root swells and accumulates large numbers of greenish oil bodies, hence the name 'pickle' (Fig. 9.6c). Cells taken from such roots form callus and somatic embryos in culture in conditions under which wild-type root cells are quiescent.

The embryonic traits observed in *pkl* seedlings are probably a consequence of the expression of *LEC1*. In *pkl* mutants, *LEC1* transcription stops as normal at the end of embryo development. However, unlike wild-type seedlings, *pkl* seedlings re-initiate *LEC1* transcription during germination. Therefore, *PKL* is required to maintain the suppression of *LEC1* transcription in seedlings. *PKL* encodes a member of the CHD3 protein family. Such proteins are known to be involved in the chromatin remodelling required for the maintenance of established patterns of transcriptional repression. Therefore, the PKL protein may directly mediate the suppression of *LEC1* transcription in the seedling.

What, then, establishes the pattern of *PKL* activity? The *PKL* gene is expressed constitutively, suggesting that regulation at the level of the PKL protein is required for the maintenance of *LEC1* suppression in the seedling. *pkl* mutants produce dark green leaves with short petioles, they flower late, are shorter than wild type, and accumulate gibberellins. These are all phenotypes associated with a reduced gibberellin response, suggesting that the PKL protein mediates gibberellin signalling. Combining *pkl* with mutations in known gibberellin synthesis or response genes has synergistic phenotypic effects, suggesting that these genes act in overlapping pathways. As described in the next case study, for example, gibberellins promote flowering. In

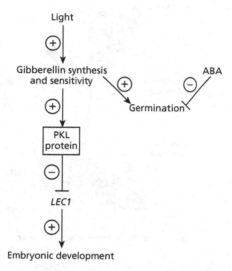

Fig. 9.7 The coordination of germination with the inhibition of embryonic development in *Arabidopsis*. Light promotes gibberellin synthesis by the seed and increases gibberellin sensitivity. Gibberellins promote germination and the activity of the PKL protein. PKL inhibits the transcription of the *LEC1* gene, thereby preventing the expression of embryonic traits by the seedling. Arrows accompanied by a 'plus' sign indicate positive regulation. Lines ending in a bar and accompanied by a 'minus' sign indicate negative regulation.

continuous light, *pkl* mutants flower about 1 day later than wild type, the gibberellin-insensitive *gai* mutant flowers about 3 days late, but *pkl gai* double mutants flower 33 days late.

These data suggest a model for the transition from embryonic to juvenile development in which gibberellins promote PKL activity, which in turn maintains the suppression of *LEC1* transcription, hence inactivating the embryonic developmental programme (Fig. 9.7).

Case study 9.3: Phase transitions in post-germination development

As described at the beginning of this chapter, the shoot apical meristem (SAM) produces a series of phytomers, each consisting of a leaf, axillary SAM and internode. The morphology of the phytomer changes as the shoot progresses through the juvenile, adult and reproductive phases. In *Arabidopsis*, internodes produced during juvenile and adult development do not expand and consequently the shoot grows as a rosette of leaves, with a very compact stem (Fig. 9.8). Rosette leaves are attached to the stem by long petioles and change gradually in shape with the progression from juvenile to adult development: early leaves usually have small, circular blades whereas later leaves have larger, more oval blades. The juvenile-to-adult transition is also marked by the transition from leaves that only have trichomes on the adaxial (upper) surface, to leaves that develop trichomes on both the adaxial and abaxial (lower) surfaces. The onset of reproductive growth is characterized by a dramatic increase in stem elongation, producing an inflorescence, or bolting stem (Fig. 9.8). The basal nodes of the bolting

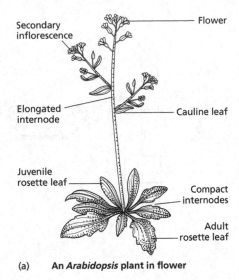

(a)   **An *Arabidopsis* plant in flower**

1 *Rosette node*

Apical meristem

Leaf primordium

No obvious axillary meristem

Base of developing rosette leaf

Older nodes further down the plant

Rosette leaf

Vegetative bud which can produce a secondary inflorescence on floral transition

2 *Basal inflorescence node*

Axillary meristem

Leaf primordium

Large axillary meristem

Base of developing cauline leaf

Cauline leaf

Secondary inflorescence

3 *Apical inflorescence node*

Axillary meristem (floral meristem)

Cryptic leaf primordium

Cryptic leaf primordium subsumed by floral meristem

Floral meristem

Sepal primordia

Flower

(b)   **The progression of leaf and axillary meristem establishment through shoot development**

Fig. 9.8 The series of phytomers produced by the *Arabidopsis* shoot. (a) An *Arabidopsis* plant in flower. During vegetative development, *Arabidopsis* produces a rosette of leaves separated by very short internodes. On floral transition, internode elongation increases rapidly and an inflorescence develops. (b) The progression of leaf and axillary meristem establishment through shoot development. 1, Rosette leaves are initiated with no obvious axillary meristem. In older nodes, away from the shoot apex, an axillary meristem develops and initiates a vegetative bud. This bud can form a secondary inflorescence on floral transition. 2, Basal nodes on the inflorescence initiate cauline leaves associated with large axillary meristems. These grow out to produce lateral inflorescences. 3, Apical nodes on the inflorescence initiate floral meristems, each of which consists of a cryptic leaf primordium and a large axillary meristem. The cryptic leaf primordium is subsumed by the floral meristem before flower development.

stem carry petioleless **cauline** leaves. Careful observation shows that some, or sometimes all, of these cauline leaves develop from leaf primordia that exist prior to floral transition. The more apical nodes of the bolting stem produce solitary flowers.

A particularly important change in phytomer morphology that accompanies phase transitions in *Arabidopsis* is a change in axillary SAM development (Fig. 9.8). In vegetative phytomers, visible axillary SAMs develop after the formation of the subtending leaf. It is not clear whether these develop *de novo* or from a small population of cells that maintain a meristematic identity throughout leaf development. Visible axillary SAMs develop in an acropetal gradient, with SAMs arising first in the axils of the oldest leaves, away from the primary SAM. The axillary SAMs of adult leaves produce small, vegetative buds. In contrast, buds visible to the naked eye are rarely produced in the axils of juvenile leaves. On the transition to floral growth, visible SAMs develop simultaneously in the axils of all leaves and grow out in a basipetal gradient, i.e. uppermost first, to produce secondary inflorescences. This means that, in contrast to adult leaves, cauline leaves have large, active axillary SAMs immediately after their inception.

Flower development involves a further increase in axillary SAM size at the expense of the subtending leaf. *Arabidopsis* flowers are relatively unusual in that they appear to lack subtending leaves. In most species, flowers in lateral positions on an inflorescence develop from axillary meristems of the leaves or bracts. In *Arabidopsis*, each floral primordium represents a leaf primordium and its axillary meristem (Fig. 9.8). The axillary meristem becomes a floral meristem and during its development subsumes the cells of the presumptive leaf. Consistent with this idea, the *AS1* gene, which is associated with leaf development (see case study 9.1), is transiently expressed at the site of the cryptic 'leaf' during flower development. In contrast to the axillary SAMs of juvenile and adult leaves, which develop as indeterminate shoots, the axillary SAM of this cryptic leaf develops as a single, terminal flower.

These dramatic changes in phytomer morphology characterize phase transitions in the *Arabidopsis* shoot. This case study considers the regulation of the juvenile-to-adult and adult-to-reproductive transitions. Of these two transitions, much more is known about the floral transition and consequently this is considered first.

Floral identity

Mutational studies indicate that three **floral meristem identity (FMI) genes** are principally required to direct a primordium to develop into a flower in *Arabidopsis*. These genes are *LEAFY* (*LFY*), *APETELA1* (*AP1*) and *CAULIFLOWER* (*CAL*). In *lfy* and *ap1* mutants, flowers are either replaced by vegetative shoots or have vegetative characteristics (Fig. 9.9). On early inflorescence nodes, *lfy* mutants produce abnormal shoot-like flowers and *ap1* mutants sometimes produce shoots in place of flowers. In both mutants, the phenotype becomes less severe at higher nodes. Towards the top of the inflorescence, *lfy* mutants produce structures that are more

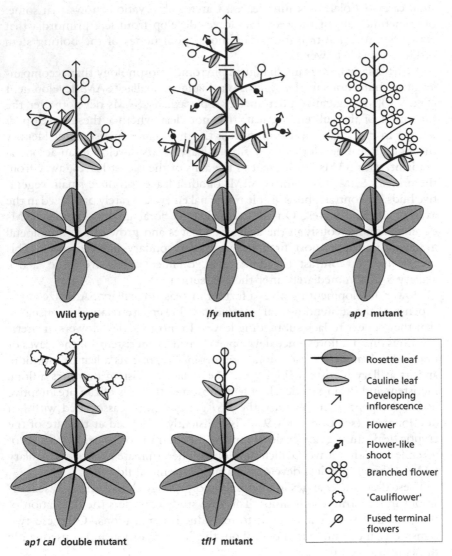

Fig. 9.9 *Arabidopsis* mutants with altered floral identity and/or inflorescence architecture. The figure shows the wild-type flowering phenotype and the phenotypes of *lfy* mutants, *ap1* mutants, *ap1 cal* double mutants and *tfl1* mutants. On *lfy* mutants, flowers are replaced by shoots or flower-like shoots (shoots with whorled rather than spiral phyllotaxy). On *ap1* mutants, flowers are sometimes replaced by shoots and flower development is abnormal: bract-like organs replace sepals, and meristems in the axils of the bracts develop into secondary flowers, producing a 'branched' flower. The phenotypes of *lfy* and *ap1* mutants become less severe at higher inflorescence nodes. In the *ap1 cal* double mutant, flowers are replaced by inflorescence meristems, which proliferate to produce miniature 'cauliflowers'. In the *tfl1* mutant, the primary inflorescence terminates in an apical flower or in a cluster of fused flowers, and secondary inflorescences are replaced by single flowers. (Adapted from Haughn *et al.*, 1995; Liljegren *et al.*, 1999.)

flower-like and *ap1* mutants produce 'branched' flowers, in which secondary flowers grow from SAMs produced in the axils of leaf-like sepals.

cal single mutants have a wild-type phenotype but *ap1 cal* double mutants fail to develop flowers. Instead, after reproductive phase change, the shoot apex of the double mutant forms cauliflower-like masses of meristematic tissue (Fig. 9.9). The miniature 'cauliflowers' arise because the inflorescence meristem, i.e. the meristem at the apex of the inflorescence (Chapter 1), initiates secondary inflorescence meristems in the place of flowers. The secondary meristems initiate tertiary meristems, and the process reiterates until a huge number of adjacent meristems have formed. Cultivated cauliflowers arise by a similar process due in part to a mutation in the cauliflower *CAL* homologue.

Since *cal* mutants have no phenotype in the presence of a functional *AP1* gene, *AP1* can entirely compensate for the loss of *CAL*. Similarly, the phenotype of *ap1* mutants is less severe than that of the *ap1 cal* double mutant, indicating that *CAL* partially compensates for the loss of *AP1*. Functional redundancy also exists between *AP1* and *LFY*. As described above, *ap1* and *lfy* mutants eventually produce flower-like structures. In contrast, *lfy ap1* double mutants show an almost complete conversion of flowers into shoots. In the single mutants, therefore, *AP1* partially compensates for the loss of *LFY*, and vice versa.

Expression and action of *LFY*, *AP1* and *CAL*

The FMI genes all encode transcription factors: *AP1* and *CAL* encode very similar MADS domain proteins, whereas *LFY* encodes an unrelated transcriptional regulator. Of the three genes, only *LFY* is transcribed in the vegetative shoot apex, where it is expressed at low levels in leaf primordia. Approximately 24 hours after an *Arabidopsis* plant is induced to flower by exposure to a long day, *LFY* transcription is initiated at the sites on the SAM from which floral primordia will emerge and *LFY* is subsequently expressed at high levels in the floral primordia (Fig. 9.10). Hence the induction of flowering accelerates and increases *LFY* expression in primordia at the shoot apex. *AP1* and *CAL* are expressed approximately 72 hours after floral transition, in the cells of visible floral primordia (Fig. 9.10). *AP1* is initially expressed only in the adaxial half of the floral primordium and its expression later spreads across the whole primordium. This supports the hypothesis that the floral primordium initially represents a leaf primordium that subtends an axillary floral meristem (see above).

The expression of *AP1* is delayed and greatly reduced in *lfy* mutants. In contrast, constitutive expression of a *LFY* transgene dramatically accelerates flowering, enhances *AP1* expression in floral primordia and causes ectopic *AP1* expression in leaf primordia. These data suggest that *LFY* induces *AP1* transcription in wild-type floral primordia. Consistent with this, the LFY transcription factor binds directly to the *AP1* promoter. *LFY* is also thought to promote *CAL* expression. Interestingly, *LFY* expression is very low in *ap1 cal* double mutants, whereas constitutive expression of an *AP1* transgene enhances *LFY* expression and causes very early flowering. These data suggest that *AP1* and *CAL* act to promote *LFY* transcription, forming a positive feedback loop that generates a high level of expression of all three genes in the floral primordium (Fig. 9.10).

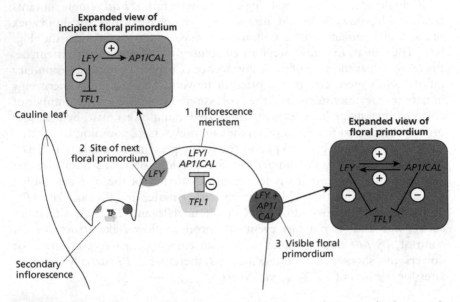

Fig. 9.10 Interactions between *LFY*, *AP1*, *CAL* and *TFL1* during inflorescence development. 1, *TFL1* is expressed just below the inflorescence meristem (pale pink) and represses *LFY*, *AP1* and *CAL* expression in the meristem, an effect probably mediated by production of an inhibitory signal (blocked line and bar). 2, *LFY* is upregulated at the site of the next floral primordium (pink) where it promotes *AP1* and *CAL* expression, and inhibits *TFL1* expression (see expanded view). 3, *LFY*, *AP1* and *CAL* are expressed in visible floral primordia (red), where they mutually promote each other's expression and continue the inhibition of *TFL1* (see expanded view). Arrows accompanied by a 'plus' sign indicate positive regulation. Lines ending in a bar and accompanied by a 'minus' sign indicate negative regulation. (Adapted from Hempel *et al.*, 2000.)

As discussed in case study 5.4, one of the main functions of the FMI genes is to initiate the transcription of the ABC genes, which control floral organ identity and without which the flower will produce leaves instead of floral organs. In this context, it is interesting to note that *AP1* also acts to specify sepal and petal identity (class A function).

Inflorescence architecture

The flowers of wild-type *Arabidopsis* plants are only produced in lateral positions on the inflorescence and inflorescence development is indeterminate. In contrast, in the inflorescence meristem of *terminal flower1* (*tfl1*) mutants, *LFY*, *AP1* and *CAL* are ectopically transcribed causing the meristem to develop into an apical flower and the inflorescence to terminate (Fig. 9.9). Similarly, lateral inflorescences on *tfl1* mutants are replaced by single flowers. These data suggest that *TFL1* is required to prevent FMI gene expression in inflorescence meristems.

TFL1 is expressed in cells just below the SAM, at low levels during vegetative development and at high levels during inflorescence development. It is probable, therefore, that *TFL1* acts to produce an as yet unidentified signal that inhibits FMI gene expression in the overlying meristem (Fig. 9.10). *TFL1* encodes a protein similar to RAF kinase inhibitor protein (RKIP), which regulates the activity of the RAF intracellular kinase signalling cascade.

Interestingly, in *lfy* mutants and *ap1 cal* double mutants, *TFL1* is ectopically transcribed in primordia initiated by the inflorescence meristem, suggesting that *LFY*, *AP1* and *CAL* repress *TFL1* expression in floral primordia (Fig. 9.10). Similarly, transgenic *Arabidopsis* plants that constitutively express either *LFY* or *AP1* have reduced *TFL1* expression and an inflorescence phenotype very similar to that of *tfl1* mutants: the primary inflorescence ends in a terminal flower, and solitary flowers replace lateral inflorescences. These observations suggest that inflorescence architecture is regulated by mutual inhibition between *TFL1* and the FMI genes (Fig. 9.10). *TFL1* specifies inflorescence meristem identity, whereas *LFY*, *AP1* and *CAL* specify floral meristem identity. In a system such as this, the final outcome can be determined simply by controlling the starting conditions. Meristem fate at any particular position will be determined by whichever antagonist, i.e. *TFL1* or the FMI genes, has the higher initial activity.

The repression of FMI genes by *TFL1* has the additional effect of delaying floral transition. In *tfl1* mutants, the numbers of adult and cauline leaves are both reduced, suggesting that the *TFL1*-dependent signal acts as a general inhibitor of the progression through successive stages of shoot development. Consistent with this model, transgenic plants with constitutive *TFL1* expression have an enlarged rosette, produce far more cauline leaves than wild type, and display a more gradual transition from lateral inflorescence to floral identity, producing a range of shoot-like flowers at intermediate nodes on the primary inflorescence. The two roles of *TFL1*—as an inhibitor of phytomer progression and as an inhibitor of terminal flower development—can be unified if the apical flower of *tfl1* mutants is considered as the final stage in the progression from vegetative to floral development, a stage that is not reached in wild-type plants. In this context, it is interesting to note that the inflorescences of many angiosperms end in a terminal flower, and this is believed to be the ancestral form of inflorescence architecture (see box 9.2).

Box 9.2: Inflorescence evolution

Arabidopsis and *Antirrhinum* both have indeterminate inflorescences. However, taxonomic studies suggest that the inflorescence of their most recent common ancestor ended in an apical flower, implying that indeterminate inflorescences evolved independently in the two species. Despite this, in *Antirrhinum*, indeterminate inflorescence development depends on a *TERMINAL FLOWER1* (*TFL1*) homologue called *CENTRORADIALIS* (*CEN*). *CEN*, like *TFL1*, is required to inhibit the expression of floral meristem identity genes in the inflorescence meristem, and in *cen* mutants the *Antirrhinum* inflorescence ends in an abnormal flower (see Fig. 5.17). Therefore, homologous genes, i.e. *CEN* and *TFL1*, were independently adopted for the same function during the evolution of indeterminate inflorescences. However, *TFL1* and *CEN* do not function identically. Unlike *TFL1*, *CEN* is not expressed during vegetative development and *cen* mutants of *Antirrhinum* flower at the normal time.

Signal integration in floral transition

Central to floral transition is the upregulation of the FMI genes. *LFY* is transcribed at low levels in leaf primordia and at high levels in floral primordia, and its expression increases abruptly if an *Arabidopsis* plant is transferred from short day to long day, florally inductive conditions. Plants exposed only to long days or only to short days show more gradual increases in *LFY* expression. The rate of increase is much higher in long days and this correlates with much earlier flowering. These data suggest that floral transition occurs when *LFY* expression reaches a threshold. Consistent with this, *lfy* mutants have an extended vegetative phase, producing more cauline leaves than wild-type plants.

As described above, constitutive expression of a *LFY* or *API* transgene causes very early flowering. Constitutive expression of a *CAL* transgene has a similar, but less dramatic, effect. The acceleration of flowering by constitutive *LFY* expression is largely negated in an *apl* mutant background, whereas early flowering due to constitutive *API* expression is retained in a *lfy* mutant background. Furthermore, as discussed above, the LFY protein directly promotes *API* transcription. Therefore, the induction of flowering by *LFY* is largely due to its promotion of *API*, and perhaps *CAL*, expression. The fact that *lfy* mutants eventually flower indicates that other mechanisms are also able to induce *API* and *CAL* transcription. These observations have led to a model in which mechanisms that regulate flowering time act by directly or indirectly controlling FMI gene expression (Fig. 9.11). This model cannot explain all aspects of floral transition. For example, since

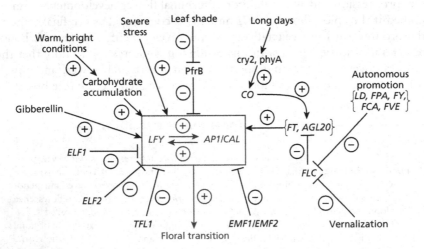

Fig. 9.11 A model of the control of floral transition in *Arabidopsis*. Floral transition is regulated by a variety of endogenous and environmental factors. The mechanisms that integrate these cues are believed to converge on the regulation of *LFY*, *API* and *CAL* expression in primordia at the shoot apex (red). *LFY*, *API* and *CAL* act to promote floral transition. Arrows accompanied by a 'plus' sign indicate positive regulation. Lines ending in a bar and accompanied by a 'minus' sign indicate negative regulation.

the FMI genes inhibit *TFL1* transcription (described above), another mechanism must be responsible for the rise in *TFL1* expression below the SAM when flowering is induced.

Environmental regulation of flowering

The effects of environmental information on flowering were discussed in the previous two chapters. In *Arabidopsis*, leaf shade (case study 7.3), long days (case study 7.5) and vernalization (case study 8.4) promote flowering (Fig. 9.11). Flowering is also accelerated by warm, bright conditions, perhaps due to increased photosynthesis. However, like many plants, *Arabidopsis* flowers early in response to severe environmental stress, such as drought, high temperature, overcrowding or nutrient deficiency.

Autonomous promotion of flowering

Even in the absence of florally promotive environmental cues, all known ecotypes of *Arabidopsis* eventually flower. This suggests that floral transition can be induced by factors endogenous to the plant. These could relate to positional, numerical or temporal information, for example the distance or number of nodes between the shoot apex and the roots, or the time since germination.

Genes required for the autonomous promotion of flowering are identified by loss-of-function mutations that cause late flowering but that do not block the environmental regulation of flowering. Examples of such genes include (in abbreviated form) *LD*, *FPA*, *FY*, *FCA* and *FVE*. Double mutants between members of this particular group have similar phenotypes to single mutants, suggesting that they form part of a single mechanism for the promotion of floral transition (Fig. 9.11). *FCA* and *LD* have been sequenced: *FCA* encodes an RNA-binding protein with protein–protein interaction domains; *LD* encodes a putative transcription factor. It is unclear how these proteins promote flowering. However, the autonomous promotion pathway as a whole most probably acts through regulation of the *FLC* gene, which inhibits flowering (see below). Mutations in most of the autonomous promotion genes result in increased *FLC* transcription, suggesting that these genes act by inhibiting *FLC* expression (Fig. 9.11). Consistent with this, such mutations do not delay flowering in *flc* null mutants.

Integration of endogenous and environmental signals

The *FLC* gene is a point of intersection between the autonomous promotion of flowering and the promotion of flowering by vernalization (Fig. 9.11). Both the autonomous mechanism (above) and vernalization (see case study 8.4) act by inhibiting *FLC* expression. *FLC* encodes a MADS domain transcription factor and acts to inhibit expression of the *AGAMOUS-LIKE 20* gene (*AGL20*, also called *SOC1*), and probably also of the *FLOWERING LOCUS T* (*FT*) gene, both of which promote flowering. As discussed in case study 7.5, the CONSTANS protein directly promotes transcription of

AGL20 and *FT* as part of the induction of flowering by long days. Therefore, *AGL20* and *FT* represent a point of intersection between the photoperiodic control of flowering, vernalization and the autonomous promotion mechanism (Fig. 9.11). Like *FLC*, *AGL20* encodes a MADS domain transcription factor. *FT* is a homologue of *TFL1* and encodes a protein similar to RKIP (see above).

Other factors that regulate floral transition

In addition to the mechanisms described above, floral transition is regulated by gibberellins and by one or more endogenous inhibition mechanisms.

Gibberellins The *ga1* mutant of *Arabidopsis*, which is severely impaired in gibberellin biosynthesis, does not flower under short days and flowers late under long days. This suggests that gibberellins are required for flowering in non-inductive photoperiods but not for the photoperiodic response. Similarly, *ga1* mutants retain a vernalization response, indicating that gibberellins are not required for the promotion of flowering by vernalization. Spraying gibberellins onto *Arabidopsis* plants promotes flowering, and *spindly* (*spy*) mutants, in which the gibberellin response pathway is constitutively active, flower early. It is probable, therefore, that an increase in gibberellin levels or response during shoot development is partly responsible for the endogenous promotion of flowering (Fig. 9.11). Gibberellins also promote the transitions from embryonic to post-embryonic development (see case study 9.2) and from juvenile to adult development (see below), suggesting that a gibberellin-mediated mechanism acts to promote all phase transitions of the shoot.

Autonomous inhibition of flowering Genes required to inhibit flowering have been identified by loss-of-function mutations that cause early flowering. Some of these mutations accelerate flowering without preventing the environmental regulation of floral transition, suggesting that the corresponding genes act in an autonomous inhibition mechanism(s) (Fig. 9.11). The most dramatic early flowering mutants of *Arabidopsis* are *embryonic flower1* and *2* (*emf1* and *emf2*). Single *emf1* or *emf2* mutants skip vegetative growth entirely and produce an inflorescence immediately after germination. The mutant inflorescence sometimes produces a few cauline leaves and mutants have highly abnormal flowers. Double *emf1 emf2* mutants lack any organized post-embryonic shoot growth. Other genes involved in the autonomous inhibition of flowering include *EARLY FLOWERING1* and *2* (*ELF1* and *ELF2*) and *TFL1*, which also regulates inflorescence architecture (see above). *elf1*, *elf2* and *tfl1* mutants flower early but after some vegetative development.

Relationship between the regulation of vegetative phase change and the regulation of flowering

In contrast to the rapidly emerging model describing the extensive and complex regulatory network that controls floral transition, very little is known about the regulation of the juvenile-to-adult transition, which is called **vegetative phase change**. The juvenile and adult phases are defined according to whether or not the shoot is competent to flower and are also characterized by the production of different modules by the SAM (see above). In most annuals, vegetative phase change occurs within weeks or even days of germination, whereas some tree species stay juvenile for one or more decades.

Many of the factors that regulate floral transition also regulate vegetative phase change. For example, vegetative phase change in *Arabidopsis* is promoted by bright light, gibberellins, long days and leaf shade. Furthermore, several mutants in which the timing of floral transition is altered display a similar change in the timing of vegetative phase change. For example, *constans* mutants lack acceleration of both vegetative and reproductive phase change by long days. Likewise, vegetative phase change is delayed in *fca* mutants, which display reduced autonomous floral promotion. Vegetative phase change may also be regulated by a temporal mechanism (see case study 3.4).

Despite these similarities, the regulation of vegetative phase change and the regulation of floral transition are not identical. In standard laboratory conditions, *Arabidopsis* produces from two to six juvenile leaves depending on the ecotype. This number is approximately the same in plants engineered to express *LFY*, *AP1* or *CAL* constitutively, even though such plants produce very few adult leaves and consequently flower very early. This suggests that the FMI genes do not promote vegetative phase change, and is consistent with the fact that shoots in the juvenile phase are not competent to flower. Interestingly, mutation of the *HASTY* (*HST*) gene causes the reciprocal phenotype to FMI gene overexpression: *hst* mutants produce fewer juvenile leaves than wild-type plants but approximately the same number of adult leaves. This suggests that *HST* acts to inhibit vegetative but not reproductive phase change. In *hst* mutants engineered to express *LFY*, both juvenile and adult leaf numbers are reduced, consistent with the hypothesis that the FMI genes and *HST* act independently.

Case study 9.4: Shoot branching

So far, this chapter has considered the initiation and maintenance of the shoot apical meristem (SAM) and the mechanisms that regulate the sequence of modules that it produces. One of the main differences in module type is the presence and fate of axillary SAMs (see case study 9.3). Vegetative axillary meristems have the potential to develop into a secondary shoot, reiterating the pattern of development of the primary shoot. This case study considers the classes of information that regulate the activity of vegetative axillary meristems and hence control the extent and pattern of shoot branching.

Factors that regulate shoot branching

In most plants each axillary SAM initiates a few leaves to form a bud, and the development of the bud then arrests. The extent of development prior to arrest varies between species, and in some plants buds can cease and then resume development several times.

The outgrowth of axillary buds into active lateral shoots is regulated by endogenous and environmental factors. Bud outgrowth is strongly promoted if the primary shoot apex is compromised, indicating that a healthy shoot apex inhibits branching (Fig. 9.12). This phenomenon is called **apical dominance**. Bud outgrowth is also regulated by the position of the bud along the primary shoot axis, and this is an important determinant of growth habit. The differences in shoot architecture between trees and shrubs, for example, occur in part because vigorous branches usually arise near to the apex of the primary shoot of young trees but near to the base of the primary shoot of shrubs. In some species, such as *Arabidopsis*, the consequences of bud position depend on whether the shoot is in the vegetative or reproductive phase. In the vegetative phase, axillary bud outgrowth in *Arabidopsis* occurs only at basal nodes, distant from the primary shoot apex. In contrast, axillary bud outgrowth in the reproductive phase occurs preferentially at apical nodes (see case study 9.3).

Environmental cues that regulate shoot branching include light quality, the gravity vector and nutrient supply. The change in light quality caused by shade from overhead foliage inhibits branching in many plants (see case study 7.3), whereas tipping a plant on its side or turning a plant upside down often promotes axillary bud outgrowth. Furthermore, it is common for buds to grow out most strongly on the upper side of a horizontally placed shoot. Nutrient sufficiency promotes shoot branching and this contrasts with its effects on roots, where nutrient sufficiency inhibits lateral root growth. It is tempting to speculate that the contrasting regulation of shoot and root branching by nutrient availability acts to balance the photosynthetic capacity of the shoot system with the nutrient-absorbing capacity of the root system (see case study 8.3).

The role of auxin

Auxin is synthesized at the shoot apex and is transported basipetally down the shoot (see case study 4.2). Hence auxin is a reliable indicator of the presence and health of the shoot apex. Consistent with this, there is strong experimental evidence that basipetal transport of apically derived auxin is required for apical dominance (Fig. 9.12). Firstly, the promotion of bud outgrowth by decapitation of the primary shoot is usually prevented if auxin is applied to the cut surface. Secondly, when auxin transport inhibitors are applied to the stem of an intact shoot, axillary bud outgrowth is often induced at nodes below the point of application. A function for auxin in the regulation of bud outgrowth is also supported by the effects of changes in endogenous auxin levels in intact plants. For example, transgenic tobacco and petunia plants with increased auxin levels have reduced branching, whereas transgenic tobacco plants with decreased auxin levels have increased branching. Similarly, *auxin resistant1* (*axr1*) mutants of *Arabidopsis*, which have reduced responses to auxin, are more branchy than wild type.

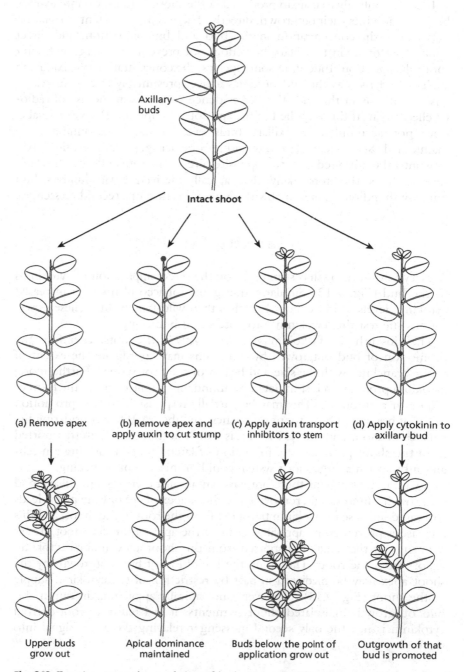

Axillary buds

Intact shoot

(a) Remove apex

(b) Remove apex and apply auxin to cut stump

(c) Apply auxin transport inhibitors to stem

(d) Apply cytokinin to axillary bud

Upper buds grow out

Apical dominance maintained

Buds below the point of application grow out

Outgrowth of that bud is promoted

Fig. 9.12 Experiments on the regulation of bud outgrowth. These are generalized results of experiments on several different species. (a) If the shoot apex is removed, one or more of the uppermost axillary buds grow out. (b) Bud outgrowth following decapitation of the shoot can be prevented if auxin is applied to the cut stump. Applying auxin directly to an axillary bud does not prevent outgrowth (not shown). (c) Applying polar auxin transport inhibitors to the stem of an intact shoot promotes the outgrowth of axillary buds below the point of application. (d) Applying cytokinin to an axillary bud promotes the outgrowth of that bud.

It is very unlikely that auxin produced at the shoot apex moves into axillary buds and inhibits their outgrowth directly. There is no consistent correlation between auxin concentration in the bud and bud inhibition, and direct application of auxin to axillary buds does not prevent bud outgrowth after shoot decapitation. Indeed, in some species, the concentration of auxin in an axillary bud rises as the bud becomes active, presumably due to increased auxin synthesis in the bud. Further evidence comes from the use of radio-labelled auxin. If this is applied to the apex of a decapitated shoot, radiolabel does not accumulate in axillary buds even though bud inhibition is maintained. Such accumulation would require acropetal (base-to-tip) auxin flow into the inhibited bud, the opposite direction to auxin flow in growing shoots. It is therefore likely that apically derived auxin inhibits bud outgrowth indirectly, presumably through the action of a second messenger.

The role of cytokinins

Axillary bud outgrowth is stimulated by the direct application of cytokinins to the bud (Fig. 9.12). Similarly, transgenic tobacco plants with increased cytokinin levels produce more branches than wild-type plants. These observations suggest that cytokinins promote shoot branching.

The biosynthesis of cytokinins occurs mostly in the roots. Therefore, the promotion of bud outgrowth by cytokinins may coordinate the extent of shoot branching with the size and health of the root system. Furthermore, cytokinin production is stimulated by abundant soil nitrogen, in the form of nitrate or ammonium. This may be partially responsible for the promotion of shoot branching by nutrient sufficiency (see above). While considering the balance of shoot and root growth, it is worth noting that auxin transported from the shoot promotes the formation of lateral roots, providing a mechanism by which a large shoot system could promote root branching.

Auxin and cytokinins therefore have antagonistic effects on axillary bud outgrowth. Interestingly, removing the shoot apex of pea or bean plants leads to a rapid increase in cytokinin transport from the root to the shoot, and this increase is prevented by applying auxin to the apex of the cut shoot. These data suggest that auxin transported from the shoot apex inhibits cytokinin export from the roots. Therefore, the inhibition of bud outgrowth by the shoot apex may be mediated in part by restriction of the cytokinin supply to the shoot (Fig. 9.13). However, since apically applied auxin can inhibit bud outgrowth in isolated shoot segments, it is likely that root-derived cytokinin is not the only second messenger relaying the auxin signal into the bud.

The *ramosus* mutants of pea

Additional components of the auxin-to-bud pathway may be defined by the *ramosus* (*rms*) mutants of pea. There are five *rms* mutants (*rms1–rms5*), all of which have increased shoot branching. The best characterized is *rms1*.

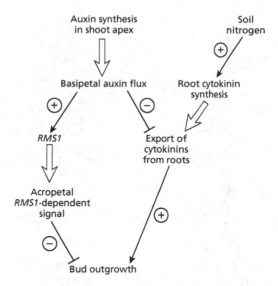

Fig. 9.13 A model for the inhibition of bud outgrowth by apically derived auxin. Auxin synthesized at the shoot apex is transported basipetally down the shoot. Basipetal auxin transport promotes the action of the *RMS1* gene, resulting in the production of an acropetally transported signal that inhibits bud outgrowth. Basipetal auxin transport also inhibits the export of cytokinins from the root. Cytokinins promote bud outgrowth when transported to the shoot. Soil nitrogen acts antagonistically to auxin, by promoting cytokinin synthesis in the root. Arrows accompanied by a 'plus' sign indicate positive regulation. Lines ending in a bar and accompanied by a 'minus' sign indicate negative regulation.

As described above, bud outgrowth following the removal of the shoot apex can be prevented if auxin is applied to the tip of the decapitated shoot. This response to applied auxin is observed in wild-type peas but not in *rms1* mutants. *rms1* mutants have wild-type auxin transport, suggesting that the *RMS1* gene acts downstream of apically derived auxin to inhibit axillary bud outgrowth (Fig. 9.13).

An elegant series of grafting experiments indicates that *RMS1* is required to produce a mobile signal that moves acropetally (Fig. 9.14). If a *rms1* scion is grafted onto a wild-type rootstock, branching in the scion is reduced to approximately wild-type levels. This suggests that the wild-type rootstock exports a signal that confers wild-type branching in the *rms1* scion. A wild-type scion grafted onto a *rms1* rootstock continues to display the wild-type branching pattern. Hence, the *RMS1* gene can confer wild-type branching when it is active only in the shoot. If a small piece of wild-type stem is grafted into the stem of a *rms1* mutant, wild-type branching is restored only to the nodes above the graft. These data suggest that *RMS1* mediates the synthesis or transport of a second messenger that relays the auxin signal back up the primary shoot and into the axillary buds (Fig. 9.13).

As described above, cytokinins promote bud outgrowth. If *RMS1* acted through the inhibition of cytokinin synthesis or transport, cytokinin export from the roots to the shoot would increase in *rms1* mutants relative to wild type. Instead, cytokinin concentration in root sap from *rms1* mutants is

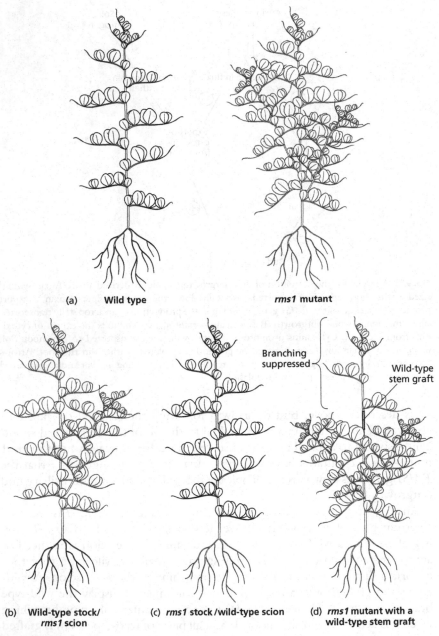

(a) **Wild type**   *rms1* mutant

Branching suppressed

Wild-type stem graft

(b) **Wild-type stock/ rms1 scion**   (c) *rms1* stock /wild-type scion   (d) *rms1* mutant with a wild-type stem graft

Fig. 9.14 Grafts between *rms1* and wild-type pea plants. (a) The *rms1* mutant (red) is more branchy than wild type. (b) If a *rms1* scion (red) is grafted onto a wild-type stock, branching in the *rms1* scion is reduced to approximately wild-type levels. (c) If a wild-type scion is grafted onto a *rms1* stock, branching in the wild-type scion remains wild type. (d) If a small section of wild-type stem (grey) is grafted into a *rms1* shoot, branching above the wild-type graft is suppressed. These results indicate that the *RMS1* gene acts in the shoot and the root and is required for the production of an acropetally transported signal that inhibits bud outgrowth.

about 15-fold lower than that in wild-type root sap, suggesting that *RMS1* action is not mediated by cytokinins. The reduction in cytokinin export from *rms1* roots correlates, as expected, with increased auxin levels (see above). Increased auxin concentrations are common in mutants with impaired auxin response, presumably reflecting negative feedback between hormone response and hormone levels.

Signal integration in the regulation of branching

The mechanism described above, in which bud outgrowth is controlled by apically derived auxin via second messengers, goes some way towards explaining the integration of endogenous and environmental factors in the regulation of branching. The auxin-to-bud pathway (Fig. 9.13) can coordinate information such as the health of the shoot apex, the relative sizes of the shoot and root systems, and the abundance of soil nutrients. However, it is not yet clear whether other regulators of branching, such as the phase of the shoot, light quality and gravity, also act through this mechanism.

Conclusions

Primary shoot development is achieved through the production of a continuous series of phytomers by the shoot apical meristem (SAM). The continuous production of phytomers is possible because development at the SAM is indeterminate. This is achieved by mechanisms that balance the proliferation of cells in the central zone of the SAM with the exit of cells to the peripheral zone and pith meristem, and also balance the proliferation of cells in these regions with organ initiation. Through the action of these mechanisms, the SAM maintains a relatively constant size and structure whilst supplying the progenitor cells for phytomer production.

Phytomers consist of a leaf, axillary SAM and internode. The structure of phytomers varies during shoot development. For example, juvenile and adult leaves often have different shapes, and axillary meristems can develop as indeterminate shoots or terminal flowers. Continuous development of a succession of phytomer types allows morphological plasticity. Developmental, physiological and environmental information is integrated to regulate how many of each phytomer type develop, and when the transitions between phytomer types occur. As well as the reiteration of phytomer production by the primary SAM, the entire shoot can be reiterated through the action of axillary SAMs. Multiple factors are integrated to regulate axillary SAM development. Existing alongside similar controls that modulate continuous and reiterative development in the root system, these mechanisms allow the plant to display an astonishing plasticity of form.

Further reading

Case study 9.1: Initiation and maintenance of the shoot apical meristem

Reviews

Barton, M.K. (1998) Cell type specification and self-renewal in the vegetative shoot apical meristem. *Current Opinion in Plant Biology* **1**, 37–42.

Bowman, J.L. & Eshed, Y. (2000) Formation and maintenance of the shoot apical meristem. *Trends in Plant Science* **5**, 110–115.

Evans, M.M.S. & Barton, M.K. (1997) Genetics of angiosperm shoot apical meristem development. *Annual Review of Plant Physiology and Plant Molecular Biology* **48**, 673–701.

Lenhard, M. & Laux, T. (1999) Shoot meristem formation and maintenance. *Current Opinion in Plant Biology* **2**, 44–50.

Surgical experiments on the SAM

For an excellent review see:

Steeves, T.A. & Sussex, I.M. (1989) Experimental investigations on the shoot apex. In: *Patterns in Plant Development*, 2nd edn, pp. 86–99. Cambridge University Press, Cambridge.

SHOOTMERISTEMLESS

Barton, M.K. & Poethig, R.S. (1993) Formation of the shoot apical meristem in *Arabidopsis thaliana*: an analysis of development in the wild type and in the *shoot meristemless* mutant. *Development* **119**, 823–831.

Long, J.A. & Barton, M.K. (1998) The development of apical embryonic pattern in *Arabidopsis*. *Development* **125**, 3027–3035.

Long, J.A., Moan, E.I., Medford, J.I. & Barton, M.K. (1996) A member of the KNOTTED class of homeodomain proteins encoded by the *STM* gene of *Arabidopsis*. *Nature* **379**, 66–69.

ASYMMETRIC LEAVES1

Byrne, M.E., Barley, R., Curtis, M. *et al.* (2000) *Asymmetric leaves1* mediates leaf patterning and stem cell function in *Arabidopsis*. *Nature* **408**, 967–971.

WUSCHEL

Laux, T., Mayer, K.F.X., Berger, J. & Jürgens, G. (1996) The *WUSCHEL* gene is required for shoot and floral meristem integrity in *Arabidopsis*. *Development* **122**, 87–96.

Mayer, K.F.X., Schoof, H., Haeker, A. *et al.* (1998) Role of *WUSCHEL* in regulating stem cell fate in the *Arabidopsis* shoot meristem. *Cell* **95**, 805–815.

Schoof, H., Lenhard, M., Haecker, A. *et al.* (2000) The stem cell population of *Arabidopsis* shoot meristems is maintained by a regulatory loop between the *CLAVATA* and *WUSCHEL* genes. *Cell* **100**, 635–644.

Partial ablation of the quiescent centre in the root apical meristem

Van den Berg, C., Willemsen, V., Hendriks, G., Weisbeek, P. & Scheres, B. (1997) Short-range control of cell differentiation in the *Arabidopsis* root meristem. *Nature* **390**, 287–289.

CLAVATA1 and CLAVATA3

Brand, U., Fletcher, J.C., Hobe, M., Meyerowitz, E.M. & Simon, R. (2000) Dependence of stem cell fate in *Arabidopsis* on a feedback loop regulated by *CLV3* activity. *Science* **289**, 617–619.

Clark, S.E., Jacobsen, S.E., Levin, J.Z. & Meyerowitz, E.M. (1996) The *CLAVATA* and *SHOOT MERISTEMLESS* loci competitively regulate meristem activity in *Arabidopsis*. *Development* 122, 1567–1575.

Clark, S.E., Williams, R.W. & Meyerowitz, E.M. (1997) The *CLAVATA1* gene encodes a putative receptor kinase that controls shoot and meristem size in *Arabidopsis*. *Cell* 89, 575–585.

Fletcher, J.C., Brand, U., Running, M.P., Simon, R. & Meyerowitz, E.M. (1999) Signaling of cell fate decisions by *CLAVATA3* in *Arabidopsis* shoot meristems. *Science* 283, 1911–1914.

Leyser, H.M.O. & Furner, I.J. (1992) Characterisation of three shoot apical meristem mutants of *Arabidopsis thaliana*. *Development* 116, 397–403.

Schoof, H., Lenhard, M., Haecker, A. *et al.* (2000) The stem cell population of *Arabidopsis* shoot meristems is maintained by a regulatory loop between the *CLAVATA* and *WUSCHEL* genes. *Cell* 100, 635–644.

Trotochaud, A.E., Jeong, S. & Clark, S.E. (2000) CLAVATA3, a multimeric ligand for the CLAVATA1 receptor-kinase. *Science* 289, 613–617.

The MGOUN genes

Laufs, P., Grandjean, O., Jonak, C., Kiêu, K. & Traas, J. (1998) Cellular parameters of the shoot apical meristem of *Arabidopsis*. *Plant Cell* 10, 1375–1389.

Case study 9.2: Transition from embryonic to post-embryonic development

LEC1

Lotan, T., Ohto, M., Yee, K.M. *et al.* (1998) *Arabidopsis* LEAFY COTYLEDON1 is sufficient to induce embryo development in vegetative cells. *Cell* 93, 1195–1205.

Meinke, D.W. (1992) A homoeotic mutant of *Arabidopsis thaliana* with leafy cotyledons. *Science* 258, 1647–1650.

West, M.A.L, Yee, K.M., Danao, J. *et al.* (1994) *LEAFY COTYLEDON1* is an essential regulator of late embryogenesis and cotyledon identity in *Arabidopsis*. *Plant Cell* 6, 1731–1745.

PICKLE

Ogas, J., Cheng, J-C., Sung, Z.R. & Somerville, C. (1997) Cellular differentiation regulated by gibberellin in the *Arabidopsis thaliana pickle* mutant. *Science* 277, 91–94.

Ogas, J., Kaufmann, S., Henderson, J. & Somerville, C. (1999) PICKLE is a CHD3 chromatin-remodeling factor that regulates the transition from embryonic to vegetative development in *Arabidopsis*. *Proceedings of the National Academy of Science, USA* 96, 13839–13844.

Case study 9.3: Phase transitions in post-embryonic development

Flowering reviews

Hempel, F.D., Welch, D.R. & Feldman, L.J. (2000) Floral induction and determination: where is flowering controlled? *Trends in Plant Science* 5, 17–21.

Koornneef, M., Alonso-Blanco, C., Peeters, A.J.M. & Soppe, W. (1998) Genetic control of flowering time in *Arabidopsis*. *Annual Review of Plant Physiology and Plant Molecular Biology* 49, 345–370.

Levy, Y.Y. & Dean, C. (1998) The transition to flowering. *Plant Cell* 10, 1973–1989.

Piñeiro, M. & Coupland, G. (1998) The control of flowering time and floral identity in *Arabidopsis*. *Plant Physiology* 117, 1–8.

Floral meristem identity genes

Ma, H. (1998) To be, or not to be a flower—control of floral meristem identity. *Trends in Genetics* 14, 26–32.

Mandel, M.A. & Yanofsky, M.F. (1995) A gene triggering flower formation in *Arabidopsis*. *Nature* 377, 522–524.

Weigel, D. & Nilsson, O. (1995) A developmental switch sufficient for flower initiation in diverse plants. *Nature* 377, 495–500.

Inflorescence architecture

Bradley, D., Carpenter, C., Copsey, L. *et al.* (1996) Control of inflorescence architecture in *Antirrhinum*. *Nature* 379, 791–797.

Bradley, D., Ratcliffe, O., Vincent, C., Carpenter, R. & Coen, E. (1997) Inflorescence commitment and architecture in *Arabidopsis*. *Science* 275, 80–83.

Liljegren, S.J., Gustafson-Brown, C., Pinyopich, A., Ditta, G.S. & Yanofsky, M.F. (1999) Interactions among *APETALA1*, *LEAFY*, and *TERMINAL FLOWER1* specify meristem fate. *Plant Cell* 11, 1007–1018.

Ratcliffe, O.J., Amaya, I., Vincent, C.A. *et al.* (1998) A common mechanism controls the life cycle and architecture of plants. *Development* 125, 1609–1615.

Ratcliffe, O.J., Bradley, D.J. & Coen, E.S. (1999) Separation of shoot and floral identity in *Arabidopsis*. *Development* 126, 1109–1120.

Late flowering mutants

Koornneef, M., Hanhart, C.J. & van der Veen, J.H. (1991) A genetic and physiological analysis of late flowering mutants in *Arabidopsis thaliana*. *Molecular and General Genetics* 229, 57–66.

FLOWERING LOCUS T

Kardailsky, I., Shukla, V.K., Ahn, J.H. *et al.* (1999) Activation tagging of the floral inducer *FT*. *Science* 286, 1962–1965.

Kobayashi, Y., Kaya, H., Goto, K., Iwabuchi, M. & Araki, T. (1999) A pair of related genes with antagonistic roles in mediating flowering signals. *Science* 286, 1960–1962.

AGAMOUS-LIKE 20

Lee, H., Suh, S-S., Park, E. *et al.* (2000) The AGAMOUS-LIKE 20 MADS domain protein integrates floral inductive pathways in *Arabidopsis*. *Genes and Development* 14, 2366–2376.

The EMF genes

Sung, Z.R., Belachew, A., Shunong, B. & Bertrand-Garcia, R. (1992) *EMF*, an *Arabidopsis* gene required for vegetative shoot development. *Science* 258, 1645–1647.

Yang, C-H., Chen, L-J. & Sung, Z.R. (1995) Genetic regulation of shoot development in *Arabidopsis*: role of the *EMF* genes. *Developmental Biology* 169, 421–435.

Vegetative phase change reviews

Kerstetter, R.A. & Poethig, R.S. (1998) The specification of leaf identity during shoot development. *Annual Review of Cell and Developmental Biology* 14, 373–398.

Lawson, E.J.R. & Poethig, R.S. (1995) Shoot development in plants: time for a change. *Trends in Genetics* 11, 263–268.

Poethig, R.S. (1990) Phase change and the regulation of shoot morphogenesis in plants. *Science* 259, 923–930.

HASTY

Telfer, A. & Poethig, R.S. (1998) *HASTY*: a gene that regulates the timing of shoot maturation in *Arabidopsis thaliana*. *Development* 125, 1889–1898.

Case study 9.4: Shoot branching

Reviews

Cline, M.G. (1991) Apical dominance. *Botanical Review* 57, 318–358.

Cline, M.G. (1994) The role of hormones in apical dominance. New approaches to an old problem in plant development. *Physiologia Plantarum* 90, 230–237.

Hobbie, L. & Estelle, M. (1994) Genetic approaches to auxin action. *Plant, Cell and Environment* 17, 525–540.

The ramosus mutants

Beveridge, C.A., Ross, J.J. & Murfet, I.C. (1996) Branching in pea. *Plant Physiology* 110, 859–865.

Beveridge, C.A., Symons, G.M., Murfet, I.C., Ross, J.J. & Rameau, C. (1997) The *rms1* mutant of pea has elevated indole-3–acetic acid levels and reduced root-sap zeatin riboside content but increased branching controlled by graft-transmissible signal(s). *Plant Physiology* 115, 1251–1258.

Beveridge, C.A., Symons, G.M. & Turnbull, C.G.N. (2000) Auxin inhibition of decapitation-induced branching is dependent on graft-transmissible signals regulated by genes *Rms1* and *Rms2*. *Plant Physiology* 123, 689–697.

Foo, E., Turnbull, C.G.N. & Beveridge, C.A. (2001) Long-distance signaling and the control of branching in the *rms1* mutant of pea. *Plant Physiology* 126, 203–209.

CHAPTER 10

A comparison of plant and animal development

Higher plants and higher animals are usually instantly distinguishable, with characteristic differences in morphology that relate to profound differences in modes and patterns of development. Such differences are not surprising since multicellularity evolved independently in plants and animals and under very different constraints from their unicellular ancestors. This chapter considers how the biology of plant and animal cells may have influenced the evolution of development, leading both to similarities and differences in the developmental mechanisms that operate in the two kingdoms. Such *post hoc* evolutionary stories are easy to construct but might bear little or no relationship to the true evolutionary constraints that led to the patterns of plant and animal development. Furthermore, extant plants and animals will by no means represent the only viable solutions to these constraints. The hypotheses set out below should therefore be treated with caution.

The most recent common ancestor of plants and animals lived over a billion years ago and was probably unicellular. It is impossible to describe this ancestor exactly but many of its characteristics have been deduced by comparing extant plant and animal cells. The common ancestor was a eukaryote that possessed mitochondria and an aerobic metabolism. Receptor proteins in its cell membrane allowed it to respond to environmental signals via the regulation of intracellular signalling networks, leading ultimately to changes in gene expression. The ancestor contained an endomembrane system, including a Golgi body, creating a compartmentalized cytoplasm and allowing endocytosis and exocytosis. It had a cytoskeleton consisting of at least actin and tubulin, allowing directed intracellular transport and the control of cell shape. Furthermore, the ancestral cell had the capacity for mitotic and meiotic division.

All of these characteristics were inherited by the progenitors of the plant and animal kingdoms. However, before multicellular plants and animals evolved, the characteristics of the two kingdoms diverged. Most importantly, the plant lineage gained chloroplasts, an autotrophic metabolism and had a cell wall. In contrast, the progenitor of multicellular animals was heterotrophic and wall-less.

Control of cell fate

The development of a zygote into a multicellular plant or animal requires the coordination of cell proliferation with precise control of cell fate. Cells must grow, differentiate, divide, migrate (in animals) or die at the correct time and place. The regulation of cell fate in both plants and animals is based on the ability of their cells to transduce intracellular and extracellular information into changes in gene activity. Plant and animal cells possess evolutionarily related signalling networks that regulate transcription; RNA processing; mRNA transport and stability; and protein modification, transport and degradation. In the unicellular ancestor, these signalling networks functioned to mediate responses to the environment and to maintain cellular homeostasis. During the evolution of multicellular plants and animals, the networks were adapted and elaborated to regulate cell fate during development. This is not a big conceptual change since in a multicellular organism, the cellular environment is largely determined by the activities of neighbouring cells.

Despite the many levels at which gene activity is regulated, it is striking, though perhaps not surprising, that cell fate in both plant and animal development often depends on the regulation of transcription. Furthermore, this frequently occurs by means of transcriptional cascades, in which cell fate is determined by the transcription of a few genes encoding transcription factors, which in turn regulate the expression of a larger number of downstream genes. For example, positional information along the anterioposterior (head-to-tail) axis of the animal embryo is transduced into position-dependent patterns of **HOX gene** expression. *HOX* genes encode a family of homeodomain transcription factors and the combination of *HOX* genes expressed by each cell regulates the combination of downstream genes that is transcribed. Mutations affecting the pattern of *HOX* gene expression cause homeotic changes in organ identity. For example, in gain-of-function *antennapedia* mutants of *Drosophila* (the fruit fly), the *HOX* gene *ANTENNAPEDIA* is ectopically expressed in the head region, causing the homeotic transformation of antennae into legs.

The ABC genes, which regulate floral organ identity, provide a similar example from plant development. As discussed in case study 5.4, all the ABC genes except for *AP2* encode MADS domain transcription factors (AP2 encodes an unrelated transcription factor) and are expressed in region-specific combinations along the radial axis of the flower. The combination of ABC genes active in a floral organ primordium specifies organ identity, and mutations in the ABC genes lead to homeotic transformations between floral organ types. For example, in the class B *ap3* mutant, petals are replaced by sepals, and stamens are replaced by carpels.

Plants and animals also share homologous mechanisms for maintaining the expression patterns of transcription factor genes, once established. For example, once the pattern of *HOX* gene transcription is established in each region of an animal, it is maintained through the action of two sets of regulatory proteins. The POLYCOMB group of proteins maintains transcriptional repression, i.e. keeps inactive *HOX* genes switched 'off', while the TRITHORAX group maintains transcriptional activation, i.e. keeps active genes switched 'on'. Interestingly, proteins with homology to the POLYCOMB group also mediate transcriptional repression in plants. One target of such repression is *AGAMOUS (AG)*, the class C floral organ identity gene in *Arabidopsis*. *AG* is expressed only in the inner two whorls of the flower, where it is required for the development of stamens and carpels. The maintenance of *AG* inactivity in the outer whorls of the flower and in the vegetative shoot requires the *CURLY LEAF (CLF)* gene, which encodes a protein homologous to a member of the POLYCOMB group.

During the evolution of multicellularity, therefore, plants and animals both adopted the transcriptional cascade as a principal mechanism for determining cell fate. Mechanisms that maintain transcriptional activity or inactivity provided a means of stabilizing cell fate. In animals, the stable inheritance of transcription between cell generations is linked to lineage restrictions on cell fate. The identity of cells as belonging to a *Drosophila* antenna, for example, can be maintained through many cell generations in culture. In contrast, plant cells are not restricted by their lineage to particular fates and will change fates if their position in a developing organ alters (Chapter 3). The mechanisms that maintain patterns of transcriptional activity in plant cells can therefore be overridden by changes in cell-extrinsic information; this is relevant to the policing of plant cell fate (discussed below).

Development of pattern

During development, cell fate is regulated to generate patterns of cells, tissues and organs. The mechanisms that generate pattern in plants and animals have several similarities. Not surprisingly, asymmetries in the egg and/or zygote often provide the information that orients and polarizes one or more of the main body axes. For example, the longitudinal axis of the *Arabidopsis* egg cell is elaborated into the shoot-to-root axis of the embryo, while the anterioposterior axis of the *Drosophila* larva is elaborated from axial information laid down in the egg by the mother.

The transition from a pattern within a single cell to a pattern across many cells can be achieved in two ways. Firstly, the original cell may divide asymmetrically into daughter cells with different fates; for example, the division of the *Arabidopsis* zygote into a terminal and basal cell (Chapter 1), or of a nematode zygote into an anterior and a posterior cell. Asymmetric divisions are common in unicellular eukaryotes, for example in the generation of spores, and this mechanism of elaborating the pattern within an individual cell probably predates the evolution of true multicellularity.

The second means of generating pattern is through intercellular signalling. Perhaps because of similar theoretical constraints, the mechanisms involved (although not the molecules) are similar in plants and animals. Commonly, this class of pattern generation entails feedback mechanisms that amplify small differences between separate regions within a field of cells to generate sharp transitions between them. The pattern of trichomes on the *Arabidopsis* leaf and that of bristles on the insect epidermis provide simple examples. In both contexts, the initial selection of 'hair' versus 'non-hair' fate is achieved by a positive feedback mechanism that amplifies the propensity of one cell in a group to form a hair, while inhibiting neighbouring cells from following suit. During subsequent growth of the leaf/epidermis, the positions of the first hairs to form determine the positions at which subsequent hairs develop.

Cell immobility versus cell mobility

Despite theoretical similarities in the mechanisms of pattern formation in plants and animals, there are significant differences. The most important of these originate from the fact that plant cells are bounded by a wall and are therefore immobile, whereas animal cells are wall-less and mobile. Cell migration is a prevalent feature of animal development and provides a major additional level at which patterns are generated. The most dramatic example is probably gastrulation, in which the animal blastula invaginates to generate the three-layered gastrula.

The mobility of animal cells also requires mechanisms that prevent inappropriate cell movement and hence the disruption of pattern. These mechanisms often depend on the regulation of the strength of adhesion between the cell and its neighbours and between the cell and the extracellular matrix. For example, the *Drosophila* wing develops from an **imaginal disc** in the larva. The wing disc is divided into overlapping anterior and posterior, and dorsal and ventral **compartments** between which cells will not migrate (Fig. 10.1). The founder cells of each compartment are specified on the basis of positional cues and compartment membership is subsequently transmitted through cell lineages. Cell migration between compartments is prevented because intercellular adhesion is greater between cells of the same compartment than between cells of different compartments.

Interestingly, the boundaries between compartments act as major signalling centres. For example, the anterioposterior pattern of cell types in the *Drosophila* wing is controlled by the concentration gradient of the DECA-PENTAPLEGIC (DPP) protein, which is produced by cells on the anterior side of the anterior–posterior compartment boundary (Fig. 10.1). Similarly, the dorsoventral pattern of cell types in the wing is regulated by the gradient of the protein WINGLESS (WG), which is produced at the dorsal–ventral boundary. Cells at the tip of the wing, i.e. at the end of the proximodistal axis, are defined by the intersection of the dorsal–ventral and anterior–posterior compartment boundaries and by the production of both DPP and WG (Fig. 10.1). The importance of compartment boundaries can be dramatically demonstrated by generating ectopic clones of 'anterior' cells in the posterior compartment of the wing. Such clones are surrounded by a new, 'anterior–posterior' boundary at which DPP is produced, and if

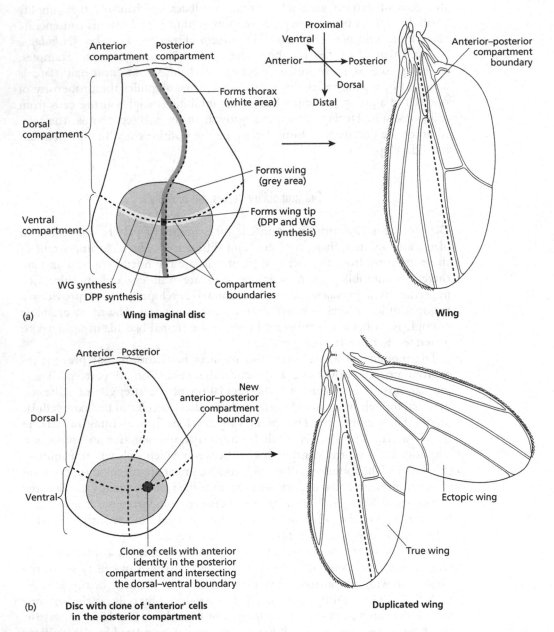

Fig.10.1 The development of the *Drosophila* wing. (a) The *Drosophila* wing develops from an imaginal disc present in the larva. The wing disc is divided into anterior and posterior, and dorsal and ventral compartments. Cells on the anterior side of the anterior–posterior boundary synthesize the morphogen DPP (pink), whereas cells at the dorsal–ventral boundary synthesize WG (pale pink). The gradients of DPP and WG concentrations across the disc direct patterning, and the intersection of DPP and WG synthesis at the junction of the two boundaries defines the wing tip (red). (b) If a clone of cells with 'anterior' identity is generated in the posterior compartment of the wing (red), a new anterior–posterior compartment boundary is formed, directing the synthesis of DPP in the 'anterior' clone. If the clone intersects the existing dorsal–ventral boundary, a second site at which both DPP and WG are synthesized is created. This site defines the tip of an ectopic wing, leading to the development of a duplicated wing. (Adapted from Tabata *et al.*, 1995; Serrano & O'Farrell, 1997.)

this intersects the natural dorsal–ventral boundary, a new wing tip is specified leading to the development of an ectopic wing (Fig. 10.1).

Although boundaries to cell migration are irrelevant to plant development, there is strong evidence that borders between cell types act as signalling centres in plants. In an interesting parallel to the development of the *Drosophila* wing, for example, the outgrowth of the leaf lamina appears to be induced by the juxtaposition of adaxial and abaxial tissue (see case study 5.1). Furthermore, mutants in which patches of ectopic abaxial epidermis develop on the adaxial leaf surface produce secondary laminas at the borders of the abaxial patches.

Policing cell fate

Having generated a pattern of cell types, multicellular organisms must police cell fates to maintain that pattern. Plants and animals achieve this through strikingly different mechanisms. Analysis of the culture conditions required by animal cells suggests that with the exception of cells in the blastula, all animal cells require **survival factors** from neighbouring cells to stay alive. In the absence of survival factors, animal cells undergo programmed cell death or **apoptosis**. The nature of the survival factors required by a particular cell depends on its cell type and, typically, the correct combination of factors are only present at the appropriate location in the organism. This therefore provides a powerful means for ensuring that cells cannot survive out of their correct context in the animal, and it is one of the safeguards that is disabled in cancer cells.

Apoptosis due to lack of survival factors is a major feature of animal development. For example, the myelin sheaths around the axons of neurons in the central nervous system are generated by oligodendrocytes. However, about twice as many oligodendrocytes are produced as are required to myelinate all the axons. This imbalance is corrected by the death of about half of the oligodendrocytes through competition for survival factors produced by the axons.

Although plant cells have the capacity for apoptosis, for example in response to infection, apoptosis is not invoked by isolation. In culture, individual plant cells not only survive, they are totipotent, able to regenerate an entire plant. This reflects the mechanism by which plant cell fate is policed. As described above, a plant cell displaced out of its normal position, will switch to a fate appropriate to its new position. Plants, therefore, reform rather than execute out-of-context cells. This mechanism depends on the maintenance of cell-extrinsic, positional information throughout the development of plant organs.

Consequences of autotrophy versus heterotrophy

The different constraints imposed by autotrophy and heterotrophy determine the range of viable plant and animal morphologies. For a photoautotrophic organism, energy and nutrients come from different sources. Hence in plants, the assimilation of energy (i.e. light) and the assimilation of nutrients and water are achieved most efficiently at separate sites, namely leaves and root hairs. Leaves also assimilate carbon dioxide. For a heterotrophic organism, energy and nutrients come from the same source (i.e. food). This imposes very different constraints on morphology. In higher animals, the mouth serves for collecting food and water; and animal morphology must allow digestion and egestion.

For plants, the production of more leaves increases the amount of sunlight and carbon dioxide absorbed, and the development of additional roots increases the uptake of water and minerals. In an environment of uniform abundance, therefore, it would be advantageous for plants to maximize their surface area. However, the real world is not uniform and often not abundant. Roots produced in a region of soil starved of nutrients, or leaves produced beneath a very dense canopy, may not repay the cost, in energy and nutrients, of their growth. Plants balance the advantages and cost of increasing surface area through indeterminate development that integrates endogenous and environmental information to control morphology.

In contrast to plants, the continuous development of new organs would not improve the feeding ability of higher animals even in conditions of abundance. Indeterminate development in animals is largely restricted to colonial invertebrates: three-headed dogs are confined to the Underworld. In most animals, morphogenesis occurs primarily during a brief embryonic phase, after which endogenous and environmental information (e.g. hunger and the sight of food) lead to changes in behaviour. In this context, it is interesting to note that the most morphologically plastic organ in an adult animal is its brain.

Conclusions

Despite separate evolution, the mechanisms that regulate plant and animal development have important similarities. These occur at the molecular level in the control of cell fate, and at the theoretical level in the generation of pattern. There are also, of course, fundamental differences in the modes of plant and animal development. It can be plausibly argued that these arose as a result of the different constraints imposed by cell immobility versus cell mobility, and by autotrophy versus heterotrophy. Of these differences, the most far reaching is probably the continuous nature of plant development compared to the determinate nature of animal development. Continuity of development allows plants to integrate endogenous and environmental information into the regulation of their morphology, leading to far greater plasticity of form than that displayed by animals.

Further reading

For discussions of animal development the reader is referred to:

Lawrence, P.A. (1992) *The Making of a Fly.* Blackwell Scientific Publications, Oxford.

Slack, J. (2000) *Essential Developmental Biology.* Blackwell Science, Oxford.

Wolpert, L., Beddington, R., Brockes, J. *et al.* (1998) *Principles of Development*. Current Biology Ltd, New York and Oxford University Press, Oxford.

For a comparison of intracellular aspects of developmental regulation, we recommend:

Meyerowitz, E.M. (1999) Plants, animals and the logic of development. *Trends in Genetics* 15, M65–M68.

Index

Printed in the United States
By Bookmasters